Titanate Based Ceramic Dielectric Materials

by R. Saravanan

Barium titanate is one of the most important electronic materials; due to its high permittivity, low dielectric loss and high tunability. The environment friendly material is suitable for microphones and microwave device applications such as tunable capacitors, delay lines, filters, resonators and phase shifters.

Doped titanates are extensively used for various electronic devices, such as transducers, piezoelectric actuators, passive memory storage devices, dynamic random access memory (DRAM), multilayer ceramic capacitors (MLCCs), positive temperature coefficient resistors (PTCR), optoelectronic devices and infrared sensors.

The book presents research results concerning the electron density distribution in a number of doped barium titanate ceramic materials using experimental X-ray diffraction data, UV-visible spectrophotometry (UV-vis), scanning electron microscopy (SEM) and energy dispersive X-ray spectroscopy (EDS). The analysis of interatomic bonding and electron density distribution is important for predicting the properties of potentially important materials and has previously been lacking for the materials studied.

Keywords: Barium Titanate, Barium Titanate Doping, Dielectric Ceramics, Permittivity, Tunability, Transducers, Piezoelectric Actuators, Memory Storage Devices, Multilayer Ceramic Capacitors, Optoelectronic Devices, X-Ray Diffraction Data, UV-Visible Spectrophotometry, Energy Dispersive X-Ray Spectroscopy, Interatomic Bonding, Electron Density Distribution, Ceramic Property Predictions.

Titanate Based Ceramic Dielectric Materials

by

Dr. R. Saravanan, M.Sc., M.Phil., Ph.D.
Associate Professor & Head
Research Centre and PG Department of Physics
The Madura College (Autonomous)
Madurai - 625 011
India

Published by **Materials Research Forum LLC**
Millersville, PA 17551, USA

Published as part of the book series
Materials Research Foundations
Volume 25 (2018)
ISSN 2471-8890 (Print)
ISSN 2471-8904 (Online)

Print ISBN 978-1-945291-54-8
ePDF ISBN 978-1-945291-55-5

This book contains information obtained from authentic and highly regarded sources. Reasonable efforts have been made to publish reliable data and information, but the author and publisher cannot assume responsibility for the validity of all materials or the consequences of their use. The authors and publishers have attempted to trace the copyright holders of all material reproduced in this publication and apologize to copyright holders if permission to publish in this form has not been obtained. If any copyright material has not been acknowledged please write and let us know so we may rectify this in any future reprints.

Distributed worldwide by

Materials Research Forum LLC
105 Springdale Lane
Millersville, PA 17551
USA
http://www.mrforum.com

Manufactured in the United States of America
10 9 8 7 6 5 4 3 2 1

Table of Contents

Preface

Barium titanate ($BaTiO_3$) has been one of the most important dielectric ceramic materials in the electronic industry. It has been the subject of various investigations related to different aspects since its discovery due to its high permittivity, low dielectric loss and high tunability. By incorporating various dopant ions in the crystal structure of $BaTiO_3$, the properties can be greatly modified. $BaTiO_3$ is a lead-free, environment friendly ceramic material which is suitable for microphones and microwave device applications such as tunable capacitors, delay lines, filters, resonators and phase shifters. Doped titanates are extensively used for various electronic devices like transducers, piezoelectric actuators, passive memory storage devices, dynamic random access memory (DRAM), multilayer ceramic capacitors (MLCCs), positive temperature coefficient resistors (PTCR), optoelectronic devices and infrared sensors. Due to a variety of significant applications of $BaTiO_3$ ceramic systems, research is recently strengthened to produce high performance titanate materials by incorporating various dopants in the crystal structure of $BaTiO_3$.

The characterization of materials is an essential part in the understanding and development of new devices and applications. The prime physical properties of any material can be clearly understood by investigating its atomic level properties. The study of precise geometrical structure, chemical bonding and electron density distribution of a material is important for predicting the properties of scientifically important materials. However, the detailed analysis of inter-atomic bonding and electron density distribution is lacking in the literature for many important materials including titanates. The present research deals with the electron density distribution studies of five series of doped barium titanate ceramic materials using experimental X-ray diffraction data. An investigation of the results obtained from other characterization works like UV-visible spectrophotometry (UV-vis), scanning electron microscopy (SEM) and energy dispersive X-ray spectroscopy (EDS) has also been carried out in this work.

Chapter 1 gives the objectives of the present research work, the importance of $BaTiO_3$ dielectric ceramic materials, description of the crystal structure of $BaTiO_3$, properties and the applications of $BaTiO_3$. The importance of doping, applications of doped titanates and the sample preparation methods of titanates for five differently doped $BaTiO_3$ are also presented. Further, this chapter discusses in detail, the characterization techniques used in this work, such as powder X-ray diffraction (PXRD), UV-visible spectrophotometry (UV-vis), scanning electron microscopy (SEM) and energy dispersive X-ray spectroscopy (EDS). In addition to the details of above experimental techniques, the analytical methodologies employed in this work namely, Rietveld method for the powder profile refinement of experimental X-ray data, maximum entropy method (MEM) for electron density analysis, Tauc's methodology for optical band gap evaluation and Scherrer formulation of grain size determination are also described elaborately.

Chapter 2 presents the results obtained from various experimental characterization and analytical techniques performed to investigate the five differently doped $BaTiO_3$ ceramic systems. The plots of experimental X-ray diffraction patterns, Rietveld fitted profiles, UV-visible absorption graphs, Tauc plots, SEM micrographs, EDS spectra, 3-dimensional (3D), 2- dimensional (2D) electron density contour maps and 1-dimensional (1D) electron density line profiles are presented in this chapter. The tables of structural parameters refined from the Rietveld method and optical band gap values with respect to the dopant concentration are also given. The elemental compositions of the prepared ceramics from EDS analysis, bond lengths and mid-bond electron density values from MEM analysis are also presented.

Chapter 3 provides the detailed analysis of results obtained for all the doped titanates from various characterization methods and analytical procedures described in chapter I. The preparation details of samples of ceramic systems chosen for this research work are presented in section 3.2. Section 3.3 gives the structural analysis of the prepared ceramic using powder X-ray diffraction, section 3.4 gives the grain size analysis and section 3.5 explains the optical characterization by UV-visible analysis. The morphological and microstructural analysis by SEM images is given in section 3.6, the chemical compositions of all the prepared ceramic systems have been analyzed in section 3.7. The precise electronic structure, inter-atomic bonding and electron density distribution in the unit cell of the prepared ceramic solid solutions are explained in section 3.8.

The interpretation of results, comparison and the correlation between the observed properties and electron density distribution for all the prepared systems are also discussed elaborately.

Chapter 4 gives the major conclusion obtained from the structural analysis, average grain size of all doped titanates, optical band gap studies, surface morphological analysis, chemical composition of the prepared ceramic solid solutions, bonding investigation and electron density distribution of various doped barium titanate ceramic materials.

Parts of the research works described in this book have been published in various Journals as follows;

1. Synthesis and analysis of electron density distribution in $Ba_{1-x}Sr_xTiO_3$ ceramics

   - R. Saravanan, J. Mangaiyarkkarasi, *Journal of Material Science: Materials in Electronics*, (Springer Publication), Vol.27, issue 3, 2523-2533, (2016)

2. Chemical bonding and charge density distribution analysis of undoped and lanthanum doped barium titanate ceramics- J. Mangaiyarkkarasi, R. Saravanan, Mukhlis M Ismail, *Journal of Chemical Science*, (Springer Publication), Vol.128, No.12, 1913-1921, (2016)

3. Effect of Ce addition on the electronic structure and bonding in $BaTi_{1-x}Ce_xO_3$ ceramics

   J. Mangaiyarkkarasi, R. Saravanan, *Journal of Material Science: Materials in Electronics*, (Springer Publication), Vol.28, issue 3, 2624-2633, (2017)

4. Electronic structure and bonding interactions in $Ba_{1-x}Sr_xZr_{0.1}Ti_{0.9}O_3$ ceramics -

   J. Mangaiyarkkarasi, S. Sasikumar, O.V. Saravanan, R. Saravanan, *Frontiers of Material Science,* (Springer Publication), Vol.11, issue 2, 182-189, (2017)

5. Charge distribution around Ba-O and Ti-O bonds in $BaTi_{1-x}Zr_xO_3$ through powder X-ray diffraction - J. Mangaiyarkkarasi, R. Saravanan, *Rare Metals*, (Springer Publication), DOI: 10.1007/s12598-016-0812-6, (Published online).

Chapter 1

Introduction

Abstract

Chapter 1 focuses on the importance of $BaTiO_3$ dielectric ceramic materials, description of the crystal structure of $BaTiO_3$, properties and the applications of $BaTiO_3$. The importance of doping, applications of doped titanates and the sample preparation methods of titanates for five differently doped $BaTiO_3$ are also presented. Further, this chapter discusses in detail, the characterization techniques used in this work, such as powder X-ray diffraction (PXRD), UV-visible spectrophotometry (UV-vis), scanning electron microscopy (SEM) and energy dispersive X-ray spectroscopy (EDS). In addition to the details of above experimental techniques, the analytical methodologies employed in this work namely, Rietveld method for the powder profile refinement of experimental X-ray data, maximum entropy method (MEM) for electron density analysis, Tauc's methodology for optical band gap evaluation and Scherrer formulation of grain size determination are also described elaborately.

Keywords

$BaTiO_3$, Ceramics, Dielectric, Charge Density, Powder Profile Refinement, Doped $BaTiO_3$

Contents

1.1   Objectives

The author synthesizes and investigates the precise electronic structure, chemical bonding, electron density distribution of five differently doped barium titanate ($BaTiO_3$) lead-free dielectric ceramic materials with various dopant compositions. Analyzed are the various effects of dopants such as, structural modification, average grain size, optical band gap variation, chemical composition and morphological changes. In order to achieve the objectives, the following tasks have been carried out.

1.   Synthesis of five different doped $BaTiO_3$ ceramic solid solutions viz.,

(i) $Ba_{1-x}Sr_xTiO_3$

(ii) $BaTi_{1-x}Zr_xO_3$

(iii) $Ba_{1-x}La_{2x/3}TiO_3$

(iv) $BaTi_{1-x}Ce_xO_3$

(v) $Ba_{1-x}Sr_xTi_{0.9}Zr_{0.1}O_3$

2.   Characterization of the prepared samples using powder X-ray diffraction (PXRD) and analysis using the Rietveld [Rietveld, 1969] refinement technique by adapting the software JANA 2006 [Petricek et al., 2014], to examine the structural properties. Evaluation of grain size using Scherrer [Cullity, 2001] formula through the GRAIN software [Saravanan, personal communication].

3.   Study of the effects of various dopants (Sr, Zr, La, Ce and Sr & Zr (co-doping)) at the host lattice sites of $BaTiO_3$ on the internal electronic structure, chemical

3

bonding and charge density distribution through the maximum entropy method (MEM) [Collins, 1982] by adapting the softwares PRIMA [Izumi, 2002] and VESTA [Momma, 2008].

4. Evaluation of the optical band gap (E_g) values of the prepared systems using UV-visible spectrophotometry (UV-vis).

5. Analysis of the microstructure and surface morphology by using scanning electron microscopy (SEM).

6. Confirmation of the elemental composition of the prepared ceramic systems qualitatively and quantitatively by energy dispersive X-ray spectroscopy (EDS).

The salient features of $BaTiO_3$, the description of the above mentioned experimental and analytical techniques and the conclusion obtained with the support of the observations are presented in the succeeding chapters.

1.2   Importance of barium titanate ($BaTiO_3$)

Recent technological improvements impose an increasing demand of the novel ceramic materials with enhanced material performance. Barium titanate ($BaTiO_3$) has been one of the highly explored ferroelectric lead-free ceramic materials since its discovery in 1940s. It has a simple crystallographic structure and has found extensive uses in the electronic and communication industry [Pradhan et al., 2013]. Barium titanate and barium titanate derived compounds have become the most promising ceramic materials to reach all the practical requirements. In $BaTiO_3$, the dielectric and ferroelectric properties arise from the covalent hybridization between Ti and O ions [Sanna et al., 2011]. $BaTiO_3$ is used as first ever piezoelectric transducers, like sensors and actuators, due to its unique ability of coupling mechanical and electrical displacements. $BaTiO_3$ derived materials offer high pressure per density ratio for actuator type devices. They are also environment friendly with chemical stability compared to other systems. The widest range of applications are, multilayer ceramic capacitors (MLCCs), piezoelectric sensors, transducers, actuators, non-volatile random access memories (NVRAM), ferroelectric random access memories (FeRAM), dynamic random access memories (DRAM), electro-optic devices etc. The positive temperature coefficient of resistance (PTCR) makes $BaTiO_3$, one of the highly wanted electro ceramic materials among the various known ceramic materials [Kumar et al., 2009].

1.3   Crystal structure of barium titanate (BaTiO₃)

Barium titanate (BaTiO$_3$) is a potential candidate of a large family of compounds with general formula ABO$_3$, termed as perovskite [Hench et al., 1990]. The perovskite structure is a primitive cube, with the larger A-cations situated at eight corners, the smaller B-cations at the body center position of the cube and the anions commonly oxygen, are at the face centers. Generally, A is a monovalent, divalent or trivalent metal atom and B is a pentavalent, tetravalent or trivalent metal atom [Vijotovic et al., 2008]. The unit cell of the ideal cubic perovskite BaTiO$_3$ is shown in figure 1.1(a). The coordination number of barium (atomic number: 56 and valency: 2+) is 12, and for titanium (atomic number: 22 and valency: 4+) it is 6.

The perovskite crystal structure can be considered as a three-dimensional network of TiO$_6$ octahedral units and this network constitutes a cubic array in which Ti-O-Ti angle is 180°. The three-dimensional network of BaTiO$_3$ structure with polyhedral representation is shown in figure 1.1(b). The undistorted ideal perovskite structure is cubic with space group $Pm\overline{3}m$ (space group number: 221). In the cubic perovskite BaTiO$_3$ structure, the barium (Ba) atom is in atomic position (0, 0, 0), the titanium (Ti) atom in (0.5, 0.5, 0.5), and the oxygen (O) atom in (0.5, 0.5, 0) [Wyckoff, 1963]. The family of perovskites not only includes compounds with cubic lattice, but also the compounds with various derivative structures from the ideal one [Hench et al., 1990].

The various principal properties of barium titanate are related to a series of structural phase transitions. The transition temperature (Curie point) T_C of BaTiO$_3$ is 120 °C. Above this Curie point and up to the temperature 1460 °C, the ideal cubic structure is stable. Above 1460 °C, the hexagonal structure is stable [Cho, 1998]. When the transition temperature is below the Curie point (from 120 °C to 5 °C), a phase transition between cubic and tetragonal structure takes place. Between 5 °C and -90 °C, the orthorhombic structure is stable, and below -90 °C, the structure is rhombohedral [Koelzynski et al., 2005]. Figures 1.2 (a) - (d) show the structural variations of BaTiO$_3$ unit cell, (a) for tetragonal (space group: $P4mm$), (b) for orthorhombic (space group: $Amm2$), (c) for rhombhohedral (space group: $R3m$) and (d) for hexagonal (space group: $P6_3/mmc$).

The stability and the formation of the ABO$_3$ perovskite based compounds have been related to the tolerance factor (t) given by Goldschmidt [Goldschmidt, 1926], which is represented as,

$$t = \frac{r_A + r_0}{\sqrt{2}(r_B + r_0)} \qquad (1.1)$$

where, r_A, r_B and r_0 are the ionic radii of the A-site, the B-site and the O-ions respectively [Choghe et al., 2004]. The tolerance factor t=1 implies the ideal cubic perovskite structure which comprises of closely packed spherical ions such that, the nearest neighbors are in contact with each other [Rick, 2009]. The deviation of the tolerance factor from the ideal value of 1 implies the distortion from the cubic structure to the tetragonal, orthorhombic or rhombhohedral type structures. When the tolerance factor t>1, the crystal structure becomes hexagonal.

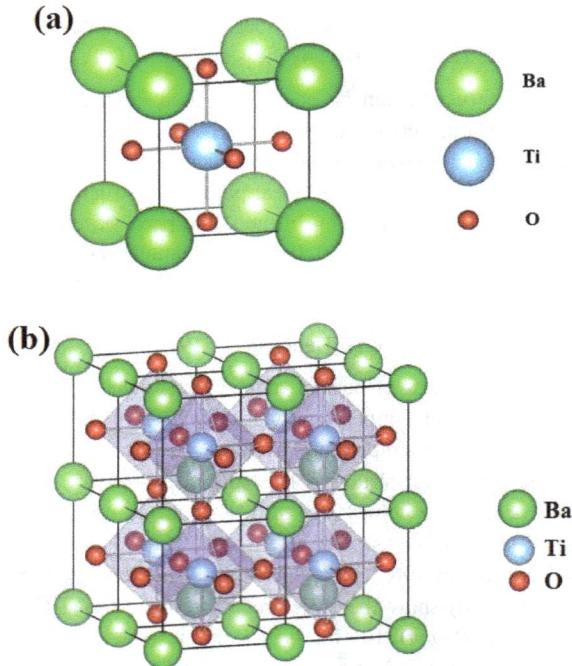

Figure 1.1 *Ideal perovskite structure of BaTiO₃ (a) cubic unit cell (space group: Pm$\bar{3}$m) (b) 3D network of perovskite structure in polyhedral representation.*

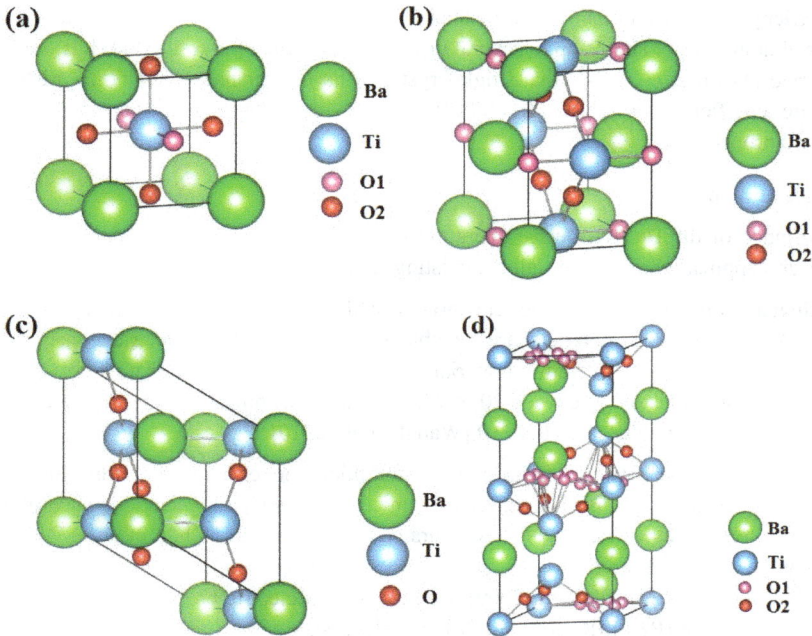

Figure 1.2 *Structural variations in BaTiO$_3$ unit cell (a) tetragonal (space group: P4mm), (b) orthorhombic (space group: Amm2), (c) rhombhohedral (space group: R3m) (d) hexagonal (space group: P6$_3$/mmc).*

1.3.1 Properties

Barium titanate, a ferroelectric perovskite is represented by the chemical formula BaTiO$_3$. It is white to grey in color when in powder form. It possesses high chemical and mechanical stability. It is soluble in sulfuric, hydrochloric and hydrofluoric acids, but insoluble in alkalis and water. The density of BaTiO$_3$ is 6.02 g/cm^3 and the melting point is 1625 °C. The transition temperature (Curie point, T$_C$) is 120 °C. BaTiO$_3$ is a perfect insulator in its pure form with a band gap of 3.2 eV, and when doped, the insulating property changes. It also exhibits semiconducting property depending on the type of dopants. However, when doped with specific amounts of isovalent or aliovalent metal ions, the properties of BaTiO$_3$ can be significantly modified. It exhibits high dielectric constant, low dielectric loss and prominent electrical tunability. Doped BaTiO$_3$ ceramic

7

materials show positive temperature coefficient of resistivity (PTCR) and piezoelectricity. At a certain temperature called Curie temperature (T_C), the resistivity of the material dramatically shoots up to several orders of magnitude [Li et al., 2005]. It was also reported [Jacob et al., 2015] that single crystal of barium titanate exhibits negative temperature coefficient of resistance (NTCR). It also exhibits optical property and pryoelectric property.

1.3.2 Uses and applications

The importance of the BaTiO$_3$ derived ceramic materials is better understandable with their uses and applications in almost all the existing fields which are listed below.

Barium titanate is a well known dielectric ceramic which is largely used for capacitors. Undoped BaTiO$_3$ ceramic materials are capable of producing the dielectric constant values as high as 7000, other ceramic materials such as titanium dioxide (TiO$_2$) possess dielectric constant values only between 20 to 70. Most of the common polymer materials have dielectric constant values less than 10 [Waugh et al., 2010].

It is a lead-free piezoceramic material suitable for microphones and other transducer applications. It is also a promising material for microwave device applications such as tunable capacitors, delay lines, filters, resonators and phase shifters [Chou et al., 2010]. The spontaneous polarization of BaTiO$_3$ single crystals at room temperature ranges from 0.15 C/m^3 to 0.26 C/m^3 [Shieh et al., 2009]. Due to its piezoelectric property, it replaced lead zirconate titanate (PZT) [Rout et al., 2011]. Polycrystalline barium titanate ceramic displays positive temperature coefficient of resistance (PTCR) which makes it suitable for thermistors and self regulating heating systems [Yoon et al., 2000].

With active optical properties such as photoluminescence and electroluminescence, the crystalline titanates find applications in new optoelectronic devices with superior performance [Melo et al., 2004]. These materials are also used in self-pumped phase conjugation (SPCC) applications due to their high reflectivity and in continuous four-wave mixing with milliwatt (mW) optical power range [Dou et al., 1996]. When combined with metals like iron, BaTiO$_3$ materials are largely employed in photorefractive type devices.

Moreover, barium titanate in the form of thin film efficiently displays electro-optic modulation of the frequencies over 40 GHz [Tang et al., 2004]. The ferroelectric and pyroelectric properties of BaTiO$_3$ are utilized in uncooled sensors for thermal cameras [Livingston et al., 2009]. Highly pure barium titanate has been reported as a key material for some capacitor energy storage systems in electric vehicles. It is also exclusively used in holographic memory systems [Funakoshi et al., 2005]. Owing to these numerous important applications of pure and doped $BaTiO_3$ ceramic systems, research is now intensified to develop $BaTiO_3$ related ceramic materials with efficient performance by substituting various isovalent and aliovalent ions with various compositions.

1.4 Doped barium titanate

Doping is the process of the intentional substitution of impurity ions into a host lattice, which leads to novel phenomena very much different from the precursors. $BaTiO_3$ perovskite, when doped with small amounts of dopants can modify its essential properties and broadening the range of potential applications [Park et al., 1997]. Perovskite structure is well known to accommodate a large number of metallic ions with wide range of substitutions at A-site or B-site or both A and B sites. These substitutions should maintain the charge balance and keep the ionic sizes within the range of particular coordination numbers [Buscalglia et al., 2000]. The variation in ionic sizes of the dopants and slight displacement of atoms lead to distortion in the crystal structure. The substitution of larger ions expands the size of the octahedral site whereas the substitution of smaller ions shrinks the size of the octahedral site. These distortions in the octahedral sites have profound effects on the physical properties of perovskite structures [Pradhan et al., 2013].

Many investigations on $BaTiO_3$ ceramic materials have reported that the isovalent and aliovalent substitutions for Ba or Ti ions lead to remarkable changes in various characteristics [Badapanda et al., 2011]. Because of the intrinsic capability of perovskite structure to host the cations with different radii, several dopants can be accommodated in the $BaTiO_3$ lattice, which can lead to changes in electrical and dielectric properties [Badapanda et al., 2011]. Deviations in the symmetry of the perovskite structure are the main reason for the various physical and chemical properties making them suitable for various technological applications [Schwartz et al., 1997]. Electrical conductivity can also be achieved when smaller multi valence B-site transition elements are substituted and exposed to different conditions [Jana et al., 2008].

1.5 Applications of doped $BaTiO_3$

1.5.1 Barium strontium titanate (BST)

Strontium doped barium titanate (BST) is one of the most widely investigated ceramic materials because of its high dielectric constant, low dielectric loss, high charge storage density and good thermal stability [Jung et al., 2008]. Recently BST has attracted immense attention due to its strong dielectric nonlinearity under the application of electric field and adjustable Curie temperature (T_C), by adjusting the Sr content over a wide range [Xu et al., 2009]. The dielectric and ferroelectric properties of BST ceramics strongly depend on the sintering conditions, grain size, porosity, amount of doping and structural defects [Deshpande et al., 2005]. The desired properties make BST a potential candidate material for tunable microwave dielectric devices [Sahoo et al., 2009].

BST ceramics are extensively used in multilayer ceramic capacitors (MLCCs), actuators, sensors, dynamic random access memory (DRAM), microwave phase shifters and transducers [Zhang et al., 2012]. Dielectric resonator antennas (DRAs) made up of BST materials have become attractive due to zero conductor loss, better gain and efficiency compared to other materials. They are effectively used in de-coupling capacitors, microwave tunable devices including tunable mixers, delay lines, filters and phase shifters for steerable antennas [Al-Zoubi et al., 2007]. They are also used in electro optic devices, piezoelectric sensors, uncooled IR detectors, pyroelectric detectors and in thermal applications [Golmohammad et al., 2010].

1.5.2 Barium zirconium titanate (BZT)

Zirconium is one of the most attractive additives for $BaTiO_3$ ceramics which strongly affects the dielectric properties. Zirconium doped barium titanate (BZT) is a promising material for lead-free actuator and for the improvement of electrostrictive properties. BZT show a very high dielectric constant because of the pinching effect. The electrical properties exhibited by this material are mainly due to the lattice expansion effect. They are widely used in the fabrication of ceramic capacitors because Zr^{4+} is more stable than Ti^{4+} [Omar et al., 2014]. Zr substitution at Ti site has been reported as an effective way to decrease the Curie temperature.

BZT single crystals have displayed piezoelectric coefficients (d_{33}) up to 850 pC/N compared to pure $BaTiO_3$ [Rehrig et al., 2000]. These excellent piezoelectric properties make Zr doped $BaTiO_3$ ceramic as a best replacement for lead-based piezoceramic materials. The substitution of Zr by Ti would depress the conduction by electronic hopping between Ti^{4+} and Ti^{3+} and thus largely decreases the leakage current which makes BZT as a suitable alternating material for BST [Sarangi et al., 1999]. Microwave dielectric properties of these BZT materials can find applications in storage capacitors for the next DRAM generation, FeRAMs and dielectric material for MLCCs, non-volatile random access memories and in microwave device applications [Cavalcante et al., 2009].

1.5.3 Barium lanthanum titanate (BLT)

$BaTiO_3$ doped with rare earth elements have been an upcoming field of research for several decades and among them lanthanum (La) has been an element of great usage. Rare earth elements are referred to as "vitamin of modern chemical industry" [Liu et al.,

2009]. A small amount of La^{3+} when substituted at Ba site leads to the formation of n-type semiconductor. If the sample is heat treated in a reducing or argon atmosphere, it becomes an insulator [Mancic et al., 2008]. BLT ceramics are largely used in positive temperature coefficient of resistor (PTCR) applications, barrier layer capacitors and inter granular capacitors [Itoh et al., 2002].

The PTCR properties of BLT make it to be used as thermistors which found large applications as follows; (i) current limiting devices for circuit protection in fuses, (ii) timers in degaussing coil circuits and (iii) in most of the CRT displays. This PTCR property is also used in heaters in the automotive industry to provide additional heat energy inside the cabin or to heat diesel in cold climatic conditions. They are widely used as temperature compensated synthesizers and also as voltage controlled oscillators. They are largely used in lithium battery protection circuits and in electrically operated wax motor to supply the heat energy required to expand the wax [Niesz et al., 2011].

Recently, researchers are developing CO gas sensors by using perovskite type semiconducting ceramic materials based on La doped $BaTiO_3$ with PTCR properties [Zhou et al., 2001]. Further, Yoo et al., [2003] have determined the p-type conductivity in the La doped barium titanate. High temperature electrical conductivity and Hall mobility studies conducted by Ali et al., [2011] have confirmed that La doped $BaTiO_3$ ceramics are very much useful for electrical and electronic applications.

1.5.4 Barium cerium titanate (BCT)

Cerium doped $BaTiO_3$ (BCT) ceramics have been extensively investigated because of their high endurance under DC electric field stress, grain growth inhibition and effective Curie temperature shift [Hwang et al., 2000, Hennings et al., 1994]. They have been used as a basic material for multilayer ceramic capacitor (MLCC) applications. The addition of cerium is able to improve the temperature stability effectively due to the good dielectric constant, enhanced dielectric properties at room temperature, low dielectric loss and large insulation resistivity.

BCT ceramic material acts as a promising material for lead free actuators and they are also employed in the wide range of current applications. These applications include self-pumped phase conjugator (SPPC), image processing, optical computing, mode locking of lasers and optical interferrometry [Qian et al., 2001]. These BCT materials are used as the best catalytic support for automatic exhaust [Rahman et al., 2014]. Ce doped $BaTiO_3$ materials are also used in optical switching devices and in real time holography systems due to their photorefractive property [Carvalho et al., 2005].

1.5.5 Barium strontium zirconium titanate (BSZT)

Simultaneous addition of Sr and Zr in the $BaTiO_3$ lattice structure is an effective way to preserve and largely improve the desirable properties. Isovalent substitution of Sr^{2+} in the place of Ba^{2+} and Zr^{4+} at the lattice site of Ti^{4+} substantially reduces the Curie temperature, which enhances the dielectric constant of the material [Manoj et al., 2008]. Adding Sr in various compositions in the $BaZr_{0.1}Ti_{0.9}O_3$ basic solid solution has shown superior ferroelectric and dielectric properties [Bhaskar et al., 2009].

BSZT compounds have been largely used as miniature microwave dielectric resonating devices in microwave integrating circuits [Fan et al., 2010]. It was reported by Jain et al., [2016] that, the BSZT ceramic materials have shown better piezoelectric constant values, overall increase of dielectric constant and negative temperature coefficient of resistivity (NTCR) behavior suggesting that these materials are best suitable for multilayer ceramic capacitor materials (MLCCs) and thermistors. The schematic diagram of MLCC is shown in figure 1.3.

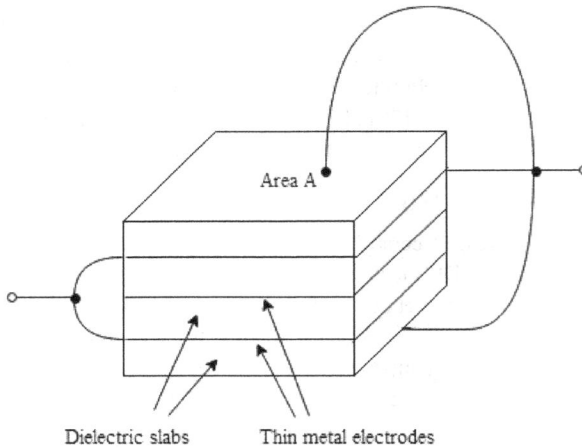

Figure 1.3 Schematic diagram of multilayer ceramic capacitor (MLCC)

1.6 Preparation of titanates

There is always an essential need to improve the useful properties of solids, which lead to the development of new essential materials to meet the present day technological innovations. Various preparation strategies are tried by many researchers to synthesize $BaTiO_3$ related ferroelectric ceramic materials. For example, the wet chemical method

[Jayanthi et al., 2004], the complex polymerization technique [Motta et al., 2008], the combustion technique [Julphunthong et al., 2013], the sonochemical method [Seeharaj et al., 2013], the microwave hydrothermal synthesis [Khanfekr et al., 2014], the citric acid gel method [Wang et al., 2007], the pechini process [Lazarevic et al., 2008], the spray pyrolysis [Choi et al., 2012], the pulsed laser deposition [Cheng et al., 2005] and the template grain growth method [Rehrig et al., 2000]. The sample density, porosity and microstructure are closely related to the method of preparation [Moulson et al., 2003].

The purity of the starting chemicals, degree of mixing, calcination and sintering temperature and their time duration, grain size, defect concentrations, pores, grain shapes, etc., largely affect the physical properties of the synthesized materials [Chen et al., 2002]. Hence, during synthesis, various factors mentioned above have to be well controlled to produce superior materials for important device applications.

1.7 Solid state reaction technique

Polycrystalline solid solutions can be successfully synthesized using the conventional solid state reaction technique by mixing two or more chemicals [Narayan, 2009]. Solid materials usually do not react at room temperature and hence in order to facilitate the reaction, they should be treated with heat at higher temperatures [Rahamam, 2003]. Highly pure, fine grained raw materials are essential to enhance the reaction rate [Moulson et al., 2003]. The basic steps needed for the solid state reaction technique are, (i) mixing, (ii) calcination, (iii) grinding, (iv) pressing / pelletizing and (v) final sintering. The flow chart which explains the various processes involved in a typical solid state reaction is given in figure 1.4.

1.7.1 Mixing

Initially, the raw materials should be mixed properly to get a homogeneous mixture. Any lack of homogeneity may affect the important properties of the synthesized material. To get better results, the action should be vigorous enough to break up any loose aggregates. The starting materials for the preparation must be highly pure to improve the reactivity [Jaffe et al., 1971]. This mixing process should be carried out for several hours which depend upon the requirement to get homogeneous and single phase structures.

1.7.2 Calcination

The next step is the solid state reaction between the raw starting materials at suitable temperature which is called calcination. Calcination process is a pre-sintering process which is very important to enhance the interaction between the constituent chemicals by improving the inter diffusion of their ions and hence, the calcination process is considered as important part of the solid state reaction technique [Narayan, 2009]. Calcination also controls the shrinkage of the materials during the sintering process.

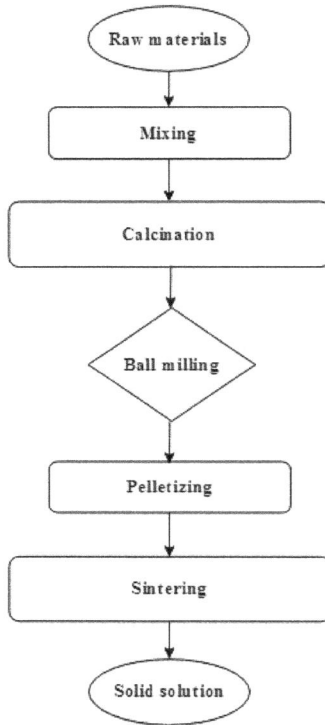

Figure 1.4 *Flow chart of a typical solid state reaction technique.*

1.7.3 Grinding

The thorough grinding of raw materials for several hours is also essential to homogenize the compositional variations, which may still exist during calcination. In laboratory practice, high energy ball milling can be used for grinding purposes. Grinding also avoids the formation of large sized particles which would be responsible for secondary grain growth during the sintering process. A ball mill works on the principle of impact and attrition. Size reduction can be achieved by the impact, as the balls drop from near the top of the sample container.

A ball mill consists of a hollow cylindrical shell, which is the sample container, made to rotate about its axis. The axis of the shell may be either horizontal or inclined at a small angle. It is partly filled with agate balls which are used to grind the sample.

When the cylindrical shell rotates, the balls are lifted upon the rising side of the shell and then they are dropped down on the sample from near the top of the shell. When this process goes on for hours, the particles are ground well and the size of the material is reduced by the impact of the balls.

1.7.4 Pressing / pelletizing

Pelletizing is necessary before final sintering, because it improves the intimate contact between the crystallites. The powder samples can be pressed into dense pellets by applying pressure using a hydraulic press. The pressure applied depends upon the composition and nature of the samples.

1.7.5 Final sintering

The final step in the solid state reaction is high temperature sintering which increases the density of the samples. The amount of oxygen during sintering needs to be controlled [Matsuo et al., 1972]. During the sintering step, the grains grow and the shape of the grain also varies significantly. Finely ground raw materials are essential to produce dense ceramics during sintering because the sintering process is connected with the surface energy of the particles. For finely ground powders, the diffusion length of the atoms involved in the sintering becomes shorter, which makes the diffusion faster, resulting in highly dense ceramics [Buckner et al., 1972, Chinor, 2000].

The standard solid state reaction technique is widely employed to prepare the titanates due to the following advantages over other methods, viz., (i) it is suitable for preparing various compositions at the same time, (ii) high temperature increases rate of diffusion, (iii) it limits the formation of side products, (iv) no waste disposal issues as in the case of solution based reactions, (v) the method is environmental friendly, (vi) no need of extensive purification to remove traces of solvent impurities as in solution based chemical methods and (vii) pelletizing improves the intimate contact between the crystallites.

In this work, the starting materials were initially mixed using an agate mortar with 12 cm in diameter. Then a laboratory ball mill, which consists of an agate jar and agate balls ranging from 1 cm and 0.6 cm diameter was utilized for grinding and mixing the raw chemicals effectively. Figure 1.5(a) shows the ball mill used in this work, which is capable of rotating in both clockwise and anticlockwise direction and figure 1.5(b) shows the agate balls. The boat type crucible made up of alumina, which withstands temperature up to 1650 °C was used in this work and it is depicted in figure 1.5(c). The ground powder samples were made as disc shaped pellets using the pelletizer which is able to provide a maximum pressure of 15 tons.

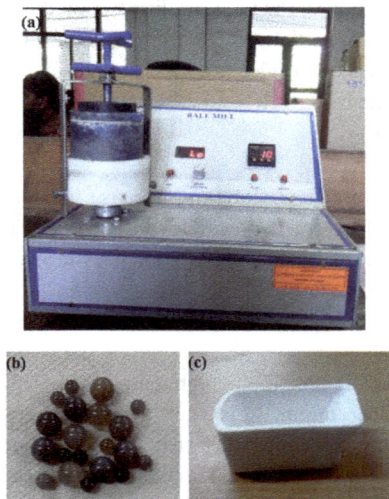

Figure 1.5 (a) *Laboratory ball mill, **(b)** agate balls, **(c)** alumina crucible.*

Figure 1.6 shows the high temperature tubular furnace equipped with Nippon/Eurotherm PID programmable controller. This furnace has the working temperature limit of up to 1600 °C with one degree accuracy of dwell temperature and has a rapid heating rate of 1 °C/min to 5 °C/min. In this work, calcinations and sintering processes were effectively carried out using the above mentioned tubular furnace.

Figure 1.6 *High temperature tubular furnace (temperature range up to 1600 °C).*

1.8 Preparation of doped titanates

In the present work, BST, BZT, BCT and BSZT solid solutions with various compositional ranges were synthesized by conventional solid state reaction route and BLT solid solution series was prepared by the chemical method. High purity (> 99.99%, Alfa aeser) starting materials were used according to the stoichiometry. The weight measurements were done using a digital balance (Model MK 200E) with a readability of 0.001 gm / 0.005 ct. Table 1.1 presents the molecular weights of the starting chemicals.

Table 1.1 *Molecular weights of chemicals used for sample preparation.*

Chemicals	Molecular weight (g/mol)
$BaCO_3$	197.35
TiO_2	79.867
$SrCO_3$	147.63
ZrO_2	123.218
CeO_2	172.115
$Ba(NO_3)_2$	261.337
$La(NO_3)_3$	324.920

1.8.1 Preparation of $Ba_{1-x}Sr_xTiO_3$

Polycrystalline $Ba_{1-x}Sr_xTiO_3$ ceramics with various Sr doping levels, x=0.2, 0.4 & 0.6 were synthesized by the standard high temperature solid state reaction technique using high purity analytical grade chemicals $BaCO_3$, $SrCO_3$ and TiO_2. The chemical reaction for the preparation is as follows;

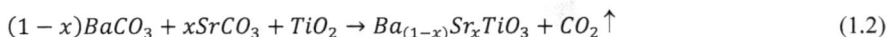

$$(1 - x)BaCO_3 + xSrCO_3 + TiO_2 \rightarrow Ba_{(1-x)}Sr_xTiO_3 + CO_2 \uparrow \qquad (1.2)$$

For the isovalent substitution of Sr^{2+} at the Ba^{2+} lattice site, the raw chemicals were weighed using a digital balance. The quantities of the raw chemicals used for the preparation of all compositions, x=0.2, 0.4 & 0.6 are given in table 1.2. The raw chemicals were thoroughly mixed using an agate mortar and mechanically activated by a high energy ball mill with the speed of 500 rpm for 5 h and calcined at 1100 °C for 10 h. Then, the powder compounds were made into dense pellets using a pelletizer by applying uniaxial pressure of 6 tons.

These pellets were placed in alumina crucibles and sintered up to 1400 °C for 5 h using a tubular furnace in air atmosphere at a heating rate of 5 °C/min. Finally, the sintered samples were cooled to room temperature and made as fine powders for characterization studies. Figures 1.7(a) - (c) show the prepared $Ba_{1-x}Sr_xTiO_3$ samples.

Table 1.2 *Quantities of chemicals used for the preparation of $Ba_{1-x}Sr_xTiO_3$.*

Sr concentration	Weight (gm)		
	$BaCO_3$	$SrCO_3$	TiO_2
0.2	1.186(4N)	0.222(4N)	0.6(4N)
0.4	0.889(4N)	0.443(4N)	0.6(4N)
0.6	0.593(4N)	0.665(4N)	0.6(4N)

The values in the parentheses indicate the purity of the chemicals.

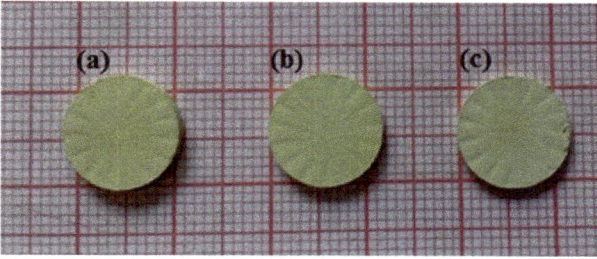

Figure 1.7 $Ba_{1-x}Sr_xTiO_3$ samples **(a)** x= 0.2, **(b)** x= 0.4 & **(c)** x= 0.6.

1.8.2 Preparation of BaTi$_{1-x}$Zr$_x$O$_3$

The lead-free BaTi$_{1-x}$Zr$_x$O$_3$ ceramics with different Zr concentrations (x=0.00, 0.04 & 0.06) were synthesized by the conventional solid state reaction technique using highly pure reagents BaCO$_3$, ZrO$_2$, and TiO$_2$. The chemical reaction for the preparation is as follows;

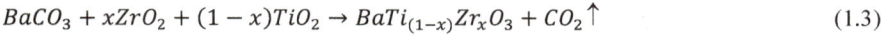

$$BaCO_3 + xZrO_2 + (1-x)TiO_2 \rightarrow BaTi_{(1-x)}Zr_xO_3 + CO_2\uparrow \qquad (1.3)$$

For the isovalent substitution of Zr^{4+} at the Ti^{4+} lattice site, the reagents were weighed with 0.001gm accuracy. The quantities of chemicals used for the preparation of BaTi$_{1-x}$Zr$_x$O$_3$ for all compositions, x= 0.00, 0.04 & 0.06 are given in table 1.3.

The stoichiometric amounts of the raw materials were thoroughly ground using an agate mortar and pestle. The mixed powder compounds were then calcined at 1200 °C for 2 h using a tubular furnace. Then, the calcined powder samples were subjected to ball milling at 200 rpm for 5 h and made into dense pellets by applying a uniaxial pressure of 6 tons. These dense pellets were sintered in air to a high temperature of 1450 °C at a rate of 5 °C/min with the dwell time of 10 h, and they were furnace cooled. Finally, the pellet samples were ground well as smooth powders for further analysis. Figures 1.8(a) - (c) show the prepared BaTi$_{1-x}$Zr$_x$O$_3$ samples.

Table 1.3 *Quantities of chemicals used for the preparation of* $BaTi_{1-x}Zr_xO_3$

Zr	Weight (gm)		
concentration	BaCO$_3$	TiO$_2$	ZrO$_2$
0.00	4.933(4N7)	1.956(4N)	-
0.04	4.933(4N7)	1.916(4N)	0.123(4N)
0.06	4.933(4N7)	1.876(4N)	0.184(4N)

The values in the parentheses indicate the purity of the chemicals

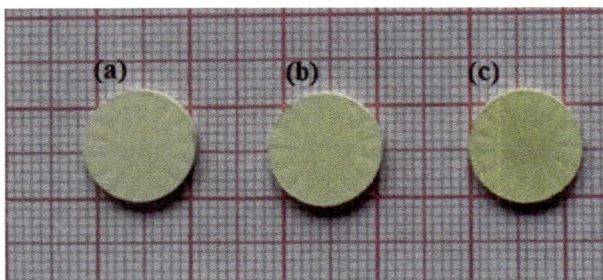

Figure 1.8 $BaTi_{1-x}Zr_xO_3$ *samples* **(a)** *x= 0.00,* **(b)** *x= 0.04 &* **(c)** *x= 0.06.*

1.8.3 Preparation of Ba$_{1-x}$ La$_{2x/3}$TiO$_3$

Rare-earth doped Ba$_{1-x}$La$_{2x/3}$TiO$_3$ (x=0.000, 0.005, 0.015, 0.020 & 0.025) samples were synthesized by the simple chemical route using highly pure Ba(NO$_3$)$_2$, La(NO$_3$)$_3$, TiO$_2$ and oxalic acid as starting chemicals. For the aliovalent substitution of La^{3+} at the site of Ba^{2+}, the quantities of chemicals used for all compositions are given in table 1.4. During the chemical synthesis, the following reaction takes place.

$$(1-x)Ba(NO_3)_2 + {^{2x}/_3}\ La(NO_3)_3 + TiO_2 \rightarrow Ba_{(1-x)}La_{2x/3}TiO_3 + NO_2 \uparrow \qquad (1.4)$$

Initially, aqueous solutions of Ba(NO$_3$)$_2$ and La(NO$_3$)$_3$ were dissolved in deionized water. Appropriate quantities of TiO$_2$ was added to oxalic acid solution and continuously stirred to form a suspension. This suspension was treated in a ultrasonic bath for 5 min to break the TiO$_2$ agglomerates. Then, Ba(NO$_3$)$_2$ and La(NO$_3$)$_3$ solutions were mixed and slowly added drop-wise into the suspension of TiO$_2$ and oxalic acid with simultaneous stirring. The pH of the resultant mixture was adjusted to 5 using ammonia solution. Finally, the resultant precipitate was filtered out and washed repeatedly using deionized water, followed by drying at 70 °C for 8 h. The dried powder samples were made into pellets and then sintered at 900 °C for 1 h. The resultant powder was ground well and utilized for further characterization studies. Figures 1.9(a) - (e) show the prepared $Ba_{1-x}La_{2x/3}TiO_3$ samples.

Table 1.4 *Quantities of chemicals used for the preparation of* $Ba_{1-x}La_{2x/3}TiO_3$

La concentration	Weight (gm)		
	Ba(NO$_3$)$_2$	La(NO$_3$)$_3$	TiO$_2$
0.000	4.355(4N)	-	1.331(4N)
0.005	4.333(4N)	0.016(4N)	1.331(4N)
0.015	4.290(4N)	0.054(4N)	1.331(4N)
0.020	4.268(4N)	0.070(4N)	1.331(4N)
0.025	4.246(4N)	0.086(4N)	1.331(4N)

The values in the parentheses indicate the purity of the chemicals

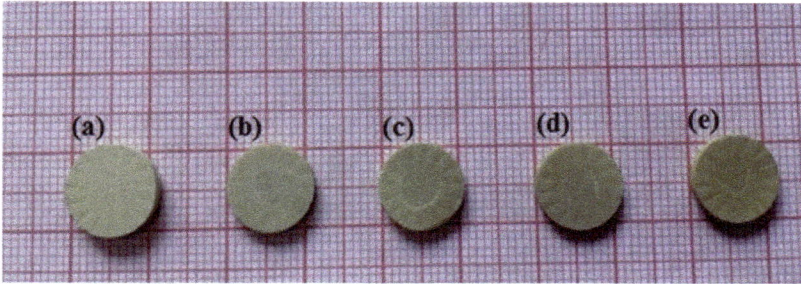

Figure 1.9 $Ba_{1-x}La_{2x/3}TiO_3$ samples *(a)* x=0.000, *(b)* x=0.005, *(c)* x=0.015, *(d)* x = 0.020 & *(e)* x=0.025.

1.8.4 Preparation of BaTi$_{1-x}$Ce$_x$O$_3$

The rare-earth substituted BaTi$_{1-x}$Ce$_x$O$_3$ ceramics with nominal Ce doping levels x=0.02, 0.04, 0.06 & 0.08 were prepared by the high temperature solid state reaction method using high purity commercial carbonates and oxides BaCO$_3$, CeO$_2$, and TiO$_2$. The chemical reaction taking place during the synthesis is given by the following equation,

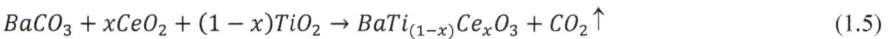

$$BaCO_3 + xCeO_2 + (1-x)TiO_2 \rightarrow BaTi_{(1-x)}Ce_xO_3 + CO_2 \uparrow \qquad (1.5)$$

For the isovalent substitution of Ce^{4+} at the lattice site of Ti^{4+}, the quantities of chemicals used in this preparation are given in table 1.5. Appropriate weights of the raw chemicals were thoroughly mixed using an agate mortar and then calcined at 1100 °C for 6 h using a tubular furnace.

The calcined powder compounds were ball milled at 200 rpm for 5 h and compressed into dense thick pellets using a pelletizer. These pellets were sintered up to a high temperature of 1400 °C for 6 h. Then, these pellets were subjected to ball milling again for another 5 h. The ball milled powder compounds were once again made into dense pellets and were sintered to 1400 °C for 12 h at a heating rate of 5 °C/min. Figures 1.10(a) - (d) show the synthesized $BaTi_{1-x}Ce_xO_3$ samples.

Table 1.5 *Quantities of chemicals used for the preparation of* $BaTi_{1-x}Ce_xO_3$

Ce	Weight (gm)		
concentration	$BaCO_3$	CeO_2	TiO_2
0.02	3.289(4N7)	0.057(4N)	1.304(4N)
0.04	3.289(4N7)	0.114(4N)	1.277(4N)
0.06	3.289(4N7)	0.172(4N)	1.251(4N)
0.08	3.289(4N7)	0.229(4N)	1.224(4N)

The values in the parentheses indicate the purity of the chemicals

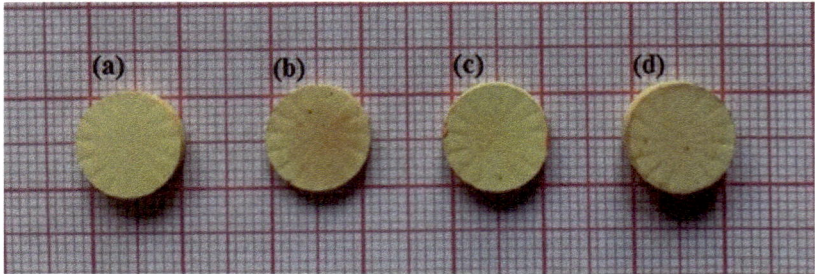

Figure 1.10 *$BaTi_{1-x}Ce_xO_3$ samples* **(a)** *x= 0.02,* **(b)** *x= 0.04,* **(c)** *x= 0.06 &* **(d)** *x= 0.08*

1.8.5 Preparation of $Ba_{1-x}Sr_xTi_{0.9}Zr_{0.1}O_3$

$Ba_{1-x}Sr_xTi_{0.9}Zr_{0.1}O_3$ ceramics with various Sr compositions x=0.00, 0.05, 0.07, & 0.14 were prepared by the conventional solid state reaction method using high purity starting materials of $BaCO_3$, $SrCO_3$, ZrO_2 and TiO_2. The following equation explains the chemical reaction taking place during the synthesis.

$$(1-x)BaCO_3 + xSrCO_3 + 0.1ZrO_2 + 0.9TiO_2 \rightarrow Ba_{1-x}Sr_xZr_{0.1}Ti_{0.9}O_3 + CO_2\uparrow \quad (1.6)$$

For the simultaneous isovalent substitution of Sr^{2+} at the lattice site of Ba^{2+} and Zr^{4+} in the place of Ti^{4+}, stoichiometric quantities of the required chemicals were calculated as given in table 1.6. These stoichiometric mixtures were thoroughly ground using an agate mortar and then calcined at 1250 °C using a tubular furnace, for 15 h.

The calcined powders were once again ground using an agate mortar and compressed into dense pellets by applying a uniaxial pressure of 6 tons using a pelletizer. These pellets were finally sintered at a temperature of 1500 °C for 8 h at a programmed heating rate of 5 °C/min. Then, the sintered samples were cooled to room temperature in the furnace and made into smooth powders for further analysis. Figures 1.11(a) - (d) show the prepared $Ba_{1-x}Sr_xTi_{0.9}Zr_{0.1}O_3$ samples.

Table 1.6 *Quantities of chemicals used for the preparation of $Ba_{1-x}Sr_xTi_{0.9}Zr_{0.1}O_3$*

Sr	Weight (gm)			
concentration	$BaCO_3$	$SrCO_3$	ZrO_2	TiO_2
0.00	3.946(4N)	-	0.246(4N)	1.437(4N)
0.05	3.749(4N)	0.147(4N)	0.246(4N)	1.437(4N)
0.07	3.670(4N)	0.206(4N)	0.246(4N)	1.437(4N)
0.14	3.354(4N)	0.442(4N)	0.246(4N)	1.437(4N)

The values in the parentheses indicate the purity of the chemicals

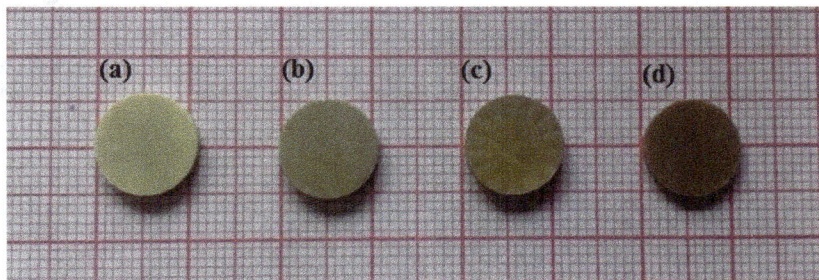

Figure 1.11 $Ba_{1-x}Sr_xTi_{0.9}Zr_{0.1}O_3$ *samples (a)* x= 0.00, *(b)* x= 0.05, *(c)* x= 0.07 & *(d)* x= 0.14.

1.9   Characterization methods

The crystal structure of a material provides a variety of important information on various scales. A proper understanding of the internal structure of the material through characterization methods has become essential to identify the novel materials for specific device applications. Currently, a wide range of experimental methods is available to investigate the mechanical, chemical or electrical properties of the chosen materials. In this work, the prepared samples have been studied using various experimental characterization techniques as given below.

(i)   Powder X-ray diffraction (PXRD)

(ii)   Ultraviolet–visible spectroscopy (UV-vis)

(iii)   Scanning electron microscopy (SEM)

(iv)   Energy dispersive X-ray spectroscopy (EDS)

The following sections briefly explain the general principles and working of the instruments involved in the characterization techniques.

1.9.1   Powder X-ray diffraction method (PXRD)

The structural information of the prepared samples can be examined through powder X-ray diffraction patterns. X-ray crystallography is a tool, which is used to identify the crystal structure of a system. By measuring the diffracting angles and intensities of the diffraction peaks, we can produce a three dimensional picture of the electron density within the crystalline system. From this electron density analysis, the mean positions of the atoms within the unit cell can be determined. Subsequently, the chemical bonds between the constituent atoms can also be analyzed.

X-ray powder diffraction (XRD) is an analytical technique primarily used for phase identification of a crystalline system, structure determination and refinement. When the monochromatic X-rays are directed towards the sample, they interact with the sample and produce constructive interference when conditions satisfy Bragg's law [Bragg, 1913] which is given by,

$$2dsin\theta = n\lambda \tag{1.7}$$

where, λ is the wavelength of monochromatic X-rays (1.54056 Å, for CuK$_\alpha$ radiation), θ is the diffraction angle, d is the lattice spacing and n is the order of the spectrum. By scanning the crystalline sample through a particular range of 2θ Bragg angles, all possible diffractions from the lattice planes can be analyzed due to the random orientation of the crystallites of the powder sample [Azaroff, 1968]. Conversion of the diffraction peaks into the lattice d-spacing facilitates the identification of the synthesized material, because each crystalline system has a set of unique d-spacings. This can be achieved by comparing the observed d-spacings with the standard reference patterns.

The schematic diagram of the X-ray diffractometer is shown in figure 1.12. The X-ray diffractometer consists of three basic components (i) X-ray tube (ii) sample holder and (iii) detector. In the X-ray tube, highly accelerated electrons are directed towards the target material. Copper is the most common target material used. When these electrons interact with the target, characteristic X-rays are produced. This X-ray spectrum consists of several components of wavelength (λ). These X-rays are then collimated to get $K_{\alpha 1}$ and $K_{\alpha 2}$ radiations and then directed towards the sample. When Bragg's law is satisfied, constructive interference occurs and the corresponding intensity peak is observed. In the θ-2θ mode spectrometer, X-ray source remains in a fixed position and the detector which is fixed in the arm moves to collect the diffraction intensities. The detector records and processes this X-ray signal and then converts the signal to a count rate [Skoog et al., 2007].

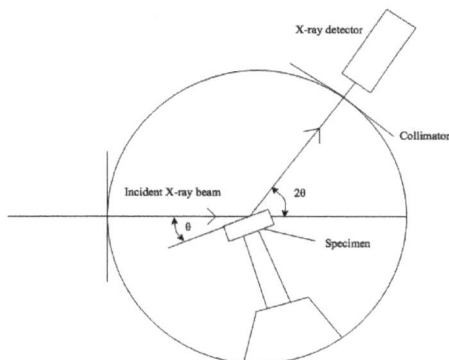

Figure 1.12 Schematic diagram of X-ray diffractometer.

1.9.2   Ultraviolet-visible spectroscopy (UV-vis)

The basic parts of a UV-visible spectrophotometer are,

    (i) light source

    (ii) monochromator

(iii) sample and reference cells and

(iv) detector.

The block diagram of UV-vis spectrometer is shown in figure 1.13. The light source is generally a tungsten filament lamp or a deuterium arc lamp. The monochromator is used to separate different wavelengths of light. The monochromator consists of prisms and slits. The radiation emitted from the light source is dispersed by the rotating prisms. After dispersion, the spectrum is focused at the exit slit. The monochromatic beam emerging out of the slit is then split into two beams.

A spectrophotometer has either single beam or double beam geometry. Nowadays double beam spectrophotometers are widely used. In double beam arrangement, the light is separated into two beams before it reaches the sample. One beam is passed through the reference and the other beam is passed through the sample. The intensity of the reference beam is taken as 100% transmission so, the intensity of radiation from the reference cell is stronger than the beam of the sample cell. This results in the generation of pulsating or alternating currents in the detector which consists of photo cells. Generally, the current generated in the photo cells is very low in intensity. So, the alternating current generated in the photocells is fed into the amplifier to amplify the signals. These signals are recorded using a pen recorder which is connected to a computer. The computer stores all the data generated and produces the spectrum of the desired sample [Gullapalli et al., 2010].

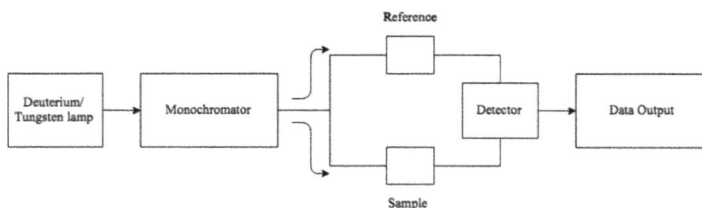

Figure 1.13 Block diagram of UV-vis spectrophotometer.

1.9.3 Scanning electron microscopy (SEM)

The schematic diagram of a scanning electron microscope (SEM) is shown in figure 1.14. The main components of the SEM instrument are (i) electron gun (ii) condenser and objective lenses (iii) scan coils (iv) sample chamber and (v) the detector. The electron gun produces a beam of electrons. The electrons ejected from the electron gun are focused towards the sample through a series of lenses. These lenses are made up of magnets capable of bending the path of the electron beam to produce clear and informative images.

The electron beam with energy in the range 0.2 KeV to 40 KeV is focused by the condenser lenses to the specimen. The electron beam, then passes through the final lens which deflects the beam in the x-axis and y-axis. The sample chamber contains the sample which is sensitive to vibrations. The deflected beam is detected by the detector [Goldstein, 2013]. SEM produces images of a sample by scanning the surface with the high energy focused beam of electrons. The incident electron beam interacts with the electrons in the sample, produce secondary electrons, back-scattered electrons and characteristic X-rays that can be detected. The normal SEM instrument operates in high vacuum to avoid the interference from the atmospheric air molecules. Magnification of SEM images can be controlled over a range of 10 to 500,000 times. The magnification control is done by adjusting the current given to the x, y deflector plates. When used with the closely related technique of energy dispersive X-ray analysis (EDS), the chemical compositions of the individual sample can also be verified.

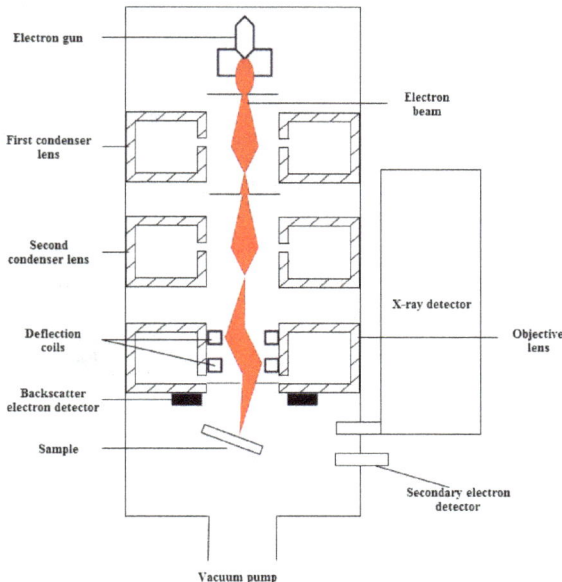

Figure 1.14 Schematic diagram of scanning electron microscope.

1.9.4 Energy dispersive X-ray spectroscopy (EDS)

The schematic diagram of energy dispersive X-ray spectrometer system is shown in figure 1.15. The principal components of a basic EDS system are (i) X-ray detector (ii) pulse processor and (iii) the computer.

When the electron beam hits the sample, X-rays will be generated. The resulting X-ray beam escapes from the sample and hits the detector, which produces a charge pulse in the detector. This short-lived current is converted into a voltage pulse with amplitude corresponding to the energy of the detected X-ray. EDS detector is a self-contained vacuum system called a cryostat with cryogenic pumping created by liquid nitrogen cooling. FET preamplifier is used to avoid the amplification of the noise signals [Agarwal, 1991].

During pulse processing, 'pulse pile-up' is possible when a second pulse reaching the amplifier before the first pulse has been fully processed. This leads to a overlapping of two pulses and thus are meaningless. Pulse pile-up rejection circuitry assumes the processing of one pulse and it is finished before accepting another. The main amplifier provides a linear, low noise amplification of the preamplification signal. All the functions of the EDS system can be fully controlled by the computer assisted systems such as multichannel analyzer (MCA). MCA contains analog to digital converter (ADC) and each received pulse is converted into a digital signal.

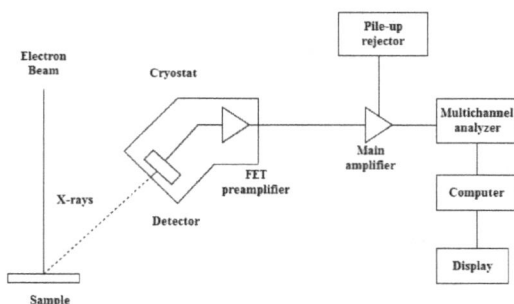

Figure 1.15 *Block diagram of an energy dispersive X-ray spectrometer (EDS) system.*

The amplitude of the pulse is proportional to the energy of the incident X-ray photon. EDS spectrum is displayed in digitized form in which, x-axis represents X-ray energy and the y-axis represents the number of counts per channel. Qualitative analysis is used for the identification elements present in the prepared samples, by analyzing the lines in the spectrum using table of energies (or) wavelengths. The quantitative analysis involves the determination of concentration of the constituent elements by measuring the peak intensities for each element in the sample.

1.10 Instruments used for characterization studies

All the XRD measurements in this work were done for the synthesized samples in their powder state. The monochromatic X-radiation (CuK_α) with wavelength 1.54056 Å was used. The powder X-ray diffraction (PXRD) patterns were obtained in the 2θ range of $10°$-$120°$ with a step size of $0.02°$ in 2θ. Two different X- ray diffractometers were used in this work.

(i)   X'pert-pro (Philips, Netherlands) at National Institute for Interdisciplinary Science and Technology (NIIST), Trivandrum, India.

(ii)   Bruker (AXS D8 advance) at Sophisticated Analytical Instrument Facility (SAIF), STIC, Cochin University, Cochin, India.

To evaluate the optical band gap (E_g), with the help of the absorption data, the prepared ceramic samples were characterized in the wavelength range 200 nm - 2000 nm by UV-visible spectrophotometer.

(i)   UV-visible spectrophotometer (Cary 5000, Varian, Germany) at Sophisticated Analytical Instrument Facility (SAIF), STIC, Cochin University, Cochin, India.

The SEM images were obtained with various magnification levels ranging from ×3000 to ×25000 using scanning electron microscopes. Two different scanning electron microscopes were utilized for the measurements.

(i)   JOEL model JSM-6390LV at Sophisticated Analytical Instrument Facility (SAIF), STIC, Cochin University, Cochin, India.

(ii)   Carl Zeiss Evo 18 at the International Research Centre, Kalasalingam University, Krishnankoil, Tamil Nadu, India.

To analyze the chemical compositions of the prepared samples, two different EDS spectrometers were utilized.

(i)   JOEL model JED-2300 at Sophisticated Analytical Instrument Facility (SAIF), Cochin University, Cochin, India

(ii)   Quantax 200 with X-Flash-Bruker at International Research Centre, Kalasalingam University, Krishnankoil, Tamil Nadu, India.

1.11 Methodologies used

In addition to the experimental characterizations discussed in the previous section, the synthesized samples were subjected to various analytical methodologies as given below.

    (i)   XRD powder profile refinement through the Rietveld [Rietveld, 1969] technique.

    (ii)   Charge density analysis through the maximum entropy method (MEM) [Collins, 1982].

    (iii)   Band gap evaluation by the Tauc plot methodology [Wood and Tauc, 1972].

    (iv)   Grain size calculation using the Scherrer [Cullity, 2001] formula.

The principles, the formulation and the procedure followed in these four methodologies are clearly explained in detail in the following sections.

1.11.1 XRD powder profile refinement through Rietveld technique

Rietveld refinement [Rietveld, 1967] is a powder profile refinement technique devised by Hugo Rietveld, who first worked out the computer based analytical procedures to make use of complete information content of the XRD powder pattern. The introduction of this technique was a significant revolution in the diffraction analysis of powder XRD patterns.

For most of the X-ray powder patterns, overlap occurs between the Bragg reflections, particularly in materials with lower symmetry due to the polycrystalline nature of the powder samples. This may lead to some loss of information of the prepared sample. The Rietveld refinement [Rietveld, 1967] technique overcomes this difficulty due to the overlapping of peaks by calculating the expected intensity for every individual step in the X-ray pattern. This method is used to analyze more complex structures by means of the least square curve fitting procedure. Rietveld refinement [Rietveld, 1967] technique minimizes the difference between the experimentally observed and theoretically built X-ray profiles. The Rietveld refinement [Rietveld, 1967] is used for refining the structural parameters related to the system and correction factors related to the instrument directly from the whole X-ray diffraction data sets.

1.11.1.1 Refinement procedure

Rietveld refinement [Rietveld, 1969] method is based on the principle of minimizing a function M which is the difference between the experimentally observed profile y(obs) and theoretically calculated profile y(cal). The function M is given as,

$$M = \sum_i W_i \left\{ y_i^{obs} - \frac{1}{c} y_i^{cal} \right\}^2 \qquad (1.8)$$

where, Σ is the sum of independent observations

W is the statistical weight

C is the overall scale factor

Based on this principle, the JANA 2006 software [Petricek et al., 2014] is employed for the profile refinement. Since the problem is not linear in the involved parameters, approximate values of all the parameters are needed in the initial refinement cycles. These are further refined in the subsequent cycles of refinement until a convergence criterion has been attained [Petricek et al., 2014]. The indicators of the quality of the Rietveld [Rietveld, 1969] refinement between the calculated and observed profiles are estimated by the residual R-factor which is given by,

$$R = \frac{\Sigma|F_{obs}-F_{cal}|}{\Sigma|F_{obs}|} \tag{1.9}$$

The R-value is used to indicate the quality of fitting between observation and constructed model. In the above expression, F_{obs} and F_{cal} are the observed and calculated structure factors. R_P and R_{obs} values indicate how well the theoretically built pattern fits with the experimental data, based on each step of the iteration. Rietveld [Rietveld, 1969] refinement uses user-selected parameters to minimize the difference between an experimental pattern and a model based on the hypothesized crystal structure and instrumental parameters. The refinable parameters used in the powder diffraction data are more than that of the single crystal structure refinement. The number of refinable parameter increases, when the prepared sample has two or more phases. The brief explanations of the parameters such as, peak shape, peak width, preferred orientation and background which are associated with the refinement are presented in the following sections.

1.11.1.2 Peak shape

The experimental profile of a single diffraction peak is empirically assumed as Gaussian due to the combined effects of X-ray beam, the experimental arrangement, the sample size and shape. The contribution of Gaussian type of distribution to the experimental profile y_i at the position $2\theta_i$ is given by,

$$y_i = I_k exp\left[-\frac{4ln(2)}{H_k^2}(2\theta_i - 2\theta_k)^2\right] \tag{1.10}$$

where, I_k is calculated intensity of the Bragg reflection, H_k is full width at half maximum (FWHM), $2\theta_k$ is the calculated position of the Bragg peak which is corrected for zero-point shift of the counter. By using the finite slit heights, together with finite sample heights, the diffraction peaks show some asymmetry at very low scattering angles. The vertical divergence effect [Klug et al., 1959] causes the maximum of the diffraction peak to shift to lower angles but not affecting the integrated peak area. In the Rietveld [Rietveld, 1969] refinement, in addition to the Gaussian shape function, the pseudo-Voigt function of Thompson, Cox & Hastings [1987] and Lorentz function are also used. A good approximation to an asymmetric peak is obtained by introducing the correction factor in equation (1.10),

$$A_s = 1 - \left[\frac{sP(2\theta_i - 2\theta_k)^2}{tan\theta_k}\right] \tag{1.11}$$

where, P is asymmetric parameter and s takes different values as, +1, 0, -1 depending on different $(2\theta_i - 2\theta_k)$ being +ve, 0, −ve respectively. Actually, at a particular position, more than one peak may contribute to the profile. So, the intensity is the sum of all the reflections contributing at the given position $2\theta_i$.

1.11.1.3 Peak width

The peak broadening which results from the particle size effect is expressed with the angular dependence of the half widths of the Bragg peaks, given by the formula [Caglioti et al., 1958],

$$H_k^2 = Utan^2\theta_k + Vtan^2\theta_k + W \tag{1.12}$$

where, U, V & W are half width parameters. The approximate values of these parameters are found by measuring the half width H_k of selected single diffraction peaks and finding a least squares fit to these observed quantities with equation (1.12).

1.11.1.4 Preferred orientation

In polycrystalline materials, at least in part of the sample, the crystallites have a tendency to align their normals along with the axis of the cylindrically shaped sample holder. When the crystallites are not in a random orientation, the intensities of reflection will vary from those of the predicted random distribution. The corrected intensity for the preferred orientation is given as,

$$I_{corr} = I_{obs} \, exp(-G\alpha^2) \tag{1.13}$$

where, α is the acute angle between the scattering vector and the normal drawn to the crystallites, G is the preferred orientation parameter and I_{obs} is the observed intensity.

1.11.1.5 Background function

The background, y_{bi}, at step i, which is approximated over a finite sum of Legendre polynomials, $F_j(x_i)$ [Abramowitz and Stegun, 1964], orthogonal relative to integration over the interval [-1, 1] is,

$$y_{bi} = \sum_{j=0}^{n} b_j F_j(x_i) \tag{1.14}$$

$F_j(x_i)$'s for $j \geq 2$ are calculated from $F_{j-1}(x_i)$ and $F_{j-2}(x_i)$ using the relation,

$$F_j(x_i) = \left(\frac{2j \pm 1}{j}\right) x_i \ F_{j \pm 1}(x_i) \pm \left(\frac{j \pm 1}{j}\right) F_{j \pm 2}(x_i) \tag{1.15}$$

with the values $F_0(x_i) = 1$ and $F_1(x_i) = x_i$. The coefficients, b_j, are called background parameters to be refined in the Rietveld [Rietveld, 1969] technique, and the variable, x_i is normalized between -1 and 1 as follows,

$$x_i = \frac{2\theta_i - \theta_{max} - \theta_{min}}{\theta_{max} - \theta_{min}} \tag{1.16}$$

The correlation coefficients between background parameters can be reduced to some extent with this background function. Even the "humps" due to amorphous or poorly crystallized material may also fit well by increasing the number of refinable background parameters.

1.11.1.6 XRD powder profile refinement using JANA 2006

In this work, the Rietveld [Rietveld, 1969] refinement was performed for all the prepared samples using the software JANA 2006 [Petricek et al., 2014]. It is a free software for structural refinement of standard, modulated and magnetic samples based on the X-ray diffraction patterns. In the Rietveld [Rietveld, 1969] refinement using JANA 2006 [Petricek et al., 2014], the observed diffraction patterns are compared with the theoretically constructed profiles using pseudo-Voigt [Wertheim, 1974], profile shape functions [Thompson, 1987] and Gaussian FWHM parameters.

The profile symmetry is also introduced by employing the Simpson rule of integration given by Howard [1982], which includes profile shape function with various coefficients and peak shift. JANA 2006 [Petricek et al., 2014] also incorporates corrections for preferred orientation using the March-Dollase function [March, 1932, Dollase, 1986]. In this way, theoretical profiles are constructed and compared with the observed profiles. The structure factors retrieved from the Rietveld [Rietveld, 1969] refinement have been utilized for the charge density studies.

1.11.2 Charge density analysis through maximum entropy method (MEM)

The aim of the present work is to completely analyze the precise electronic structure of the five differently doped $BaTiO_3$ systems and the bonding interaction between the constituent atoms. This aim is achieved successfully by adapting the maximum entropy method (MEM) [Collins, 1982]. In this section, the importance of electron density studies in structure analysis, the formalism of MEM method [Collins, 1982], the principle, MEM [Collins, 1982] methodology followed in this work to elucidate the electron density are discussed elaborately.

1.11.2.1 Electron density

The electron density is a quantum mechanical observable, that can be measured through scattering experiments, in particular, X-ray diffraction from the crystals. The possibility of measuring charge density in a crystalline system from its X-ray diffraction pattern was conceived several years ago by Debye and Scherrer [Debye et al., 1918].

Electron density is the measure of probability of an electron being present at a specific location. The regions of electron density are usually found around the atoms and the bonds between them. The probability of locating an electron at one point or another can be calculated quantum mechanically. This calculation gives a quantity called electron density. Electron can be viewed as a stationary wave or a cloud of negative charges. The electron density is considered as a periodic function of position in a crystal, reaching to a maximum value at a point where an atom is present and dropping to a minimum value in the region between the two atoms [Cullity and Stock, 2001]. An electron's wave nature is described mathematically by its orbital or wave function, and this wave function assigns the number to each point in the space, and the numbers are positive at some locations and negative at some other locations. The electron density function ρ is equal to ψ^2 which means that ρ is always positive. Electron density value expresses the relative probability of finding an electron at a particular location. An electron can also be described as a wave function or orbital. The orbital, whether atomic or molecular, covers a region of space and does not move. A moving electron looks like a stationary cloud of charges. The physical interpretation of the electron density function $\rho(r)$ is that ρdr is the probability of finding an electron in a volume element dr.

According to the quantum theory of atoms in molecules, the atoms and bonds are considered as the principal objects of the crystal structure. The electron density distribution of a molecule is the probability distribution that clearly describes the manner in which the electronic charges are distributed throughout the real space in the field exerted by the nuclei. For the detailed investigation of bonding, the electron density distribution analysis is essential [Coppens, 1989]. The most commonly adapted techniques of electron density analysis are the estimation of electron density maps and least squares fitting of parameterized analytical functions to the observed structure factors. The 3-dimensional and 2-dimensional maps are obtained by refined structure factors. The knowledge of various phases present in the observed X-ray patterns is essential for the electron density analysis [Coppens, 1979].

1.11.2.2 Structure factor

The calculation of structure factors is essential to derive the charge density inside the unit cell because the charge density is the Fourier transform of structure factors. The structure factor F_{hkl} is the resultant of j waves scattered in the direction of reflection hkl by j atoms in the unit cell [Stout and Jensen, 1968]. The expression for the structure factor is,

$$F_{hkl} = F_{hkl}exp(i\alpha_{hkl}) = \sum_j f_j exp\left[2\pi i\left(hx_j + ky_j + lz_j\right)\right] \qquad (1.17)$$

$$= \sum_j f_j cos\left[2\pi\left(hx_j + ky_j + lz_j\right)\right] + i\sum_j f_j sin\left[2\pi\left(hx_j + ky_j + lz_j\right)\right]. \qquad (1.18)$$

$$= A_{hkl} + iB_{hkl} \qquad (1.19)$$

where, the sum is taken over all atoms in the unit cell, x_j, y_j, z_j are the atomic coordinates of the jth atom, f_j is the scattering factor of the jth atom and hkl is the phase of the diffracted beam. The scattering factor f_j is the ratio between the amplitude of radiation scattered from the atom and the amplitude of radiation scattered from the single electron.

The structure factor describes the way in which an incident X-ray is scattered by all the atoms in the unit cell by considering the various scattering power of the elements through the term f_j. Due to the spatial distribution of the atoms in the unit cell, there will be a phase difference in the scattering amplitudes from the two atoms. This phase difference is taken into account by complex exponential form. When the summation over discrete atoms in the structure factor expression is,

$$F_{hkl} = \sum f_j \, e^{2\pi i(hx_j + ky_j + lz_j)} \tag{1.20}$$

In the above equation, the summation is replaced by an integration of a continuous, cyclic electron density function, ρ, and an expression is obtained that is in the form of Fourier transform,

$$F_{hkl} = \int_V \rho(x, y, z) e^{2\pi i(hx_j + ky_j + lz_j)} dV \tag{1.21}$$

This means that electron density is the Fourier transform of structure factor. Similarly, the structure factor is the inverse Fourier transform of electron density.

$$\rho(x, y, z) = \int_V F_{hkl} e^{-2\pi i(hx + ky + lz)} dV \tag{1.22}$$

The above equation can be used in the form of a summation which is written as,

$$\rho(x, y, z) = \sum_h \sum_k \sum_l F_{hkl} e^{-2\pi i(hx + ky + lz)} dV \tag{1.23}$$

According to the above equation (1.23), the electron density can be calculated at any point (x, y, z) by constructing a Fourier series which has coefficients those are equal to the structure factors [Warren, 1990]. This is the basic equation of crystallography and it enables the calculation of electron density in the unit cell.

1.11.2.3 Fourier method

We can imagine that the unit cell is divided into small volumes dV in which there are $\rho(r)dV$ number of electrons. The amplitude of the scattered wave from such a small volume will be $\rho(r)dV$ times as much that of an electron at the same position. From this fact, we find the total scattered amplitude from the electron density distribution $\rho(r)$. $F(H)$ can be written in terms of density $\rho(r)$ as,

$$F(H) = \int \rho(r) \, exp(2\pi i H . r) \, dV \tag{1.24}$$

The inverse Fourier transform of the above expression gives the electron density as,

$$\rho(r) = \int F(H)\, exp(-2\pi i H.r)\, dV = \frac{1}{V} F(H) exp(-2\pi i H.r) \tag{1.25}$$

Since F(H) is defined as the discrete set of reciprocal lattice points k, the integral can be replaced by the summation. Writing the structure factor as,

$F(H) = A(H) + iB(H)$, then,

$$\rho(r) = \frac{1}{V}\sum(A(H) + iB(H))[cos(2\pi H.r) - i\, sin(2\pi i H.r)] \tag{1.26}$$

where, $A(H)=A(-H)$ and $B(H)=B(-H)$ since the electron density is a real function.

$$\rho(r) = \frac{1}{V}2A(H)\, cos(2\pi H.r) + 2B\, sin(2\pi H.r) \tag{1.27}$$

with $A(H) = |F(H)|\, cos\,\varphi$ and $B(H) = |F(H)|\, sin\,\varphi$ resulting in,

$$\rho(r) = \frac{2}{V}\sum_{1/2}[|F(H)|\, cos\,\varphi\, cos(2\pi H.r) + |F(H)|\, sin\,\varphi\, sin(2\pi H.r)]\ldots\ldots \tag{1.28}$$

which reduces to,

$$\rho(r) = \frac{2}{V}\sum_{1/2}[|F(H)|\, cos(2\pi H.r) - \varphi(H)] \tag{1.29}$$

Each structure factor contributes a plane wave to the total density with wave vector H and phase φ. The formation of the image, which is the density, needs the understanding of the phases of the structure factors. Once an approximation to the scattering density is known, φ(H) may be calculated on the basis of this approximation, and an admittedly imperfect image of the structure can be retrieved. Anomalous scattering should also be corrected and this can be achieved by subtracting the calculated contributions $\Delta A_{calc}^{anamalous}$ and $\Delta B_{calc}^{anamalous}$ from A and B, respectively, using the anomalous scattering factors f' and f''.

The period of the plane wave with the amplitude $F(H)$, in the direction of the wave vector H is equal to 1/H. So, the period is shorter for higher-order reflections which add resolution to the image. When more higher-order reflections are included in the summation, the resolution of the image improves substantially. The improvement is similar to the enhancement of resolution in an optical image obtained with shorter wavelength radiation.

The non-existence of lenses for X-ray beams makes it necessary to adapt computational methods to achieve the Fourier transform of the diffraction pattern into the image. Infinite number of Fourier coefficients is needed to perform the Fourier Synthesis in the calculation of charge density by this methodology. But, we use only a limited number of Fourier coefficients by ignoring the experimental errors and setting all the missing Fourier coefficients as zero. In this procedure, the missing structure factors are neglected by setting them to zero because the experiment cannot be or was not carried out. This assumption is a highly biased one and this will result in a negative electron density. So, this will not provide the accurate charge density properties. Hence, in the present work, precise charge density studies have been performed by adapting the maximum entropy method which is called as MEM [Collins, 1982].

1.11.2.4 Formalism of maximum entropy method (MEM)

The maximum entropy method (MEM) [Collins, 1982] can be used to determine the electron density in the unit cell from the phased X-ray diffraction data. MEM method [Collins, 1982] is used to extract the maximum amount of information from the X-ray data [Smaalen et al., 2009]. At first, MEM [Collins, 1982] image technique was employed in radio astronomy. Later, it found applications in image processing, the analysis of any type of spectroscopic or diffraction data. MEM [Collins, 1982] is now vastly used in crystallographic applications and to determine most probable electron density distribution in the unit cell of a particular system under investigation, and locating atoms and the determination of accurate charge density and chemical bonding.

Precise electron density maps can be obtained using the MEM [Collins, 1982] method. The resolution of the MEM [Collins, 1982] map is relatively higher than the map drawn by the conventional Fourier transform [Sakata et al., 1990]. In X-ray diffraction, the experimental structure factors applied to the inverse Fourier transform yield the charge density. But, it is difficult to obtain all the observed structure factors without measurement related errors. These limitations are overcome by the MEM method [Collins, 1982] of inferring the unobserved structure factors from the experimentally observed structure factors and maximizing the entropy [Saravanan et al., 2012]. The overall intensities of each reflection are evaluated from the experimentally observed X-ray diffraction patterns using the results retrieved from the Rietveld [Rietveld, 1969] refinement. By combining the Rietveld [Rietveld, 1969] refinement and MEM method [Collins, 1982], a new sophisticated technique of structure refinement in the charge density studies was evolved [Takata et al., 1995]. This Rietveld [Rietveld, 1969] and

MEM [Collins, 1982] combinational analysis is basically an iterative procedure which can finally provide a better structural model [Takata et al., 2001]

1.11.2.5 Principle of MEM

The principle of MEM [Collins, 1982] is to obtain accurate electron density, which is consistent with the experimental structure factors and to leave the uncertainties to a minimum.

The theory of maximum entropy method (MEM) [Collins, 1982] is understandable with some equations which are similar to the equations in statistical thermodynamics because the information entropy and statistical entropy deal with the concept of most probable distribution. In the statistical thermodynamics, the distribution of the particles in phase space is considered whereas in the information theory, the distribution of numerical quantities over the ensemble of pixels is considered. The probability distribution of N identical particles over m boxes each containing n_i particles can be written as,

$$P = \frac{N!}{n_1! n_2! n_3! .. n_m!}$$

(1.30)

According to statistical thermodynamics, the entropy is defined as $\ln P$. Since the numerator in the above equation is constant, the entropy is written as,

$$S = -\sum_i n_i \ln n_i$$

(1.31)

where, Stirlings' formula $\ln N! \approx N \ln N - N$ has been used.

When there is a prior probability q_i for the ith box to contain n_i number of particles. This can be expressed by,

$$P = \frac{N!}{n_1! n_2! n_3! .. n_m!} q_1^{n_1} q_2^{n_2} ... q_m^{n_m}$$

(1.32)

which then gives the expression for entropy,

$$S = -\sum_i n_i \ln n_i + \sum_i n_i \ln q_i = -\sum_{i=1}^{m} n_i \ln \frac{n_i}{q_i}$$

(1.33)

By using the equation (1.33), the MEM method was first introduced by Collins [1982] which is expressed in information entropy of the electron density distribution as the sum over M grid points in the unit cell with the help of the entropy formula [Jaynes, 1968].

$$S = -\sum \rho'(r)\ln\left(\frac{\rho'(r)}{\tau'(r)}\right)$$

$$(1.34)$$

The probability $\rho'(r)$ and prior probability $\tau'(r)$ are related to the electron density inside the unit cell as,

$$\rho'(r) = \frac{\rho(r)}{\sum_r \rho(r)} \quad \text{and} \quad \tau'(r) = \frac{\tau(r)}{\sum_r \tau(r)}$$

$$(1.35)$$

where, $\rho(r)$ is the electron density at a certain fixed r in an unit cell and $\tau(r)$ is the prior electron density. In this formulation, the actual electron densities are treated instead of normalized densities and $\rho'(r)$ became $\tau'(r)$ when there is no information. The $\rho'(r)$ and $\tau'(r)$ are normalized as,

$$\sum \rho'(r) = 1 \quad \text{and} \quad \sum \tau'(r) = 1$$

$$(1.36)$$

Then, the entropy is maximized with a constraint, which is given by,

$$C = \frac{1}{N}\sum \frac{|F_{cal}(K) - F_{obs}(K)|^2}{\sigma^2(K)}$$

$$(1.37)$$

Here, N is the number of reflections accounted for MEM [Collins, 1982] analysis, $\sigma(k)$ is the standard deviation of the observation and calculated structural factor $F_{cal}(k)$ is expressed as,

$$F_{cal}(k) = V\sum \rho(r)\exp(-2\pi i k.r)\,dV$$

$$(1.38)$$

where, V is the volume of the unit cell.

The constraint C which is expressed in equation (1.37) is sometimes termed as a weak constraint in which the calculated structure factors agree well with the observed ones when C becomes unity. The structure factor given in equation (1.38) is the Fourier transform of the electron density inside the unit cell. The structure factors are normally written without introducing the atomic form factors. It should be stressed here that it would be an assumption to use the atomic form factors in the formulation of structure factors.

The equation (1.38) confirms that it is possible to allow any type of deformation in the electron densities in real space as long as information concerning such a deformation is included in the observed data. In order to constrain the function C to be unity during entropy maximization, the Lagrange's method of undetermined multiplier is used. Then,

$$Q = S - \left(\frac{\lambda}{2}\right) C = -\sum \rho'(r) \; ln \left(\frac{\rho'(r)}{\tau'(r)}\right) - \frac{\lambda}{2N} \sum_k \frac{|F_{cal}(k) - F_{obs}(k)|^2}{\sigma^2(k)} \tag{1.39}$$

When $\dfrac{dQ}{d\rho} = 0$ and using the approximation, $ln \; x = x\text{-}1$ then the electron density,

$$\rho(r_i) = \tau(r_i) \; exp\{(\frac{\lambda F_{000}}{N})\left[\sum \frac{1}{\sigma(k)^2}\right]|F_{obs}(k)\text{-}F_{cal}(k)|exp\,(\text{-}2\pi j \; k.r)\} \; \ldots\ldots\ldots \tag{1.40}$$

where, $F_{000} = Z$, the total number of electrons in a unit cell. Equation (1.40) cannot be solved as it is, since $F_{obs}(k)$ is defined on $\rho(r)$. To solve equation (1.40) in a easier route, the following approximation which replaces $F_{cal}(k)$ is used

$$F_{cal}(k) = V \sum \tau(r) \exp(-2\pi i k.r) dV \tag{1.41}$$

The above approximation is called zero[th] order single pixel approximation, and by using this approximation, the right hand side of equation (1.41) becomes independent of prior density $\tau(r)$ and equation (1.41) can be solved in an iterative way starting from a given density of the prior distribution. The uniform density distribution is given as the prior density $\tau(r)$ as,

$$0 \le \tau(\mathbf{r}) \ge \frac{Z}{M} \tag{1.42}$$

where, M is the number of pixels for which the electron density is calculated. The reason for the choice of prior distribution is that the uniform density distribution corresponds to the maximum entropy state among all possible density distributions. The validity of zero[th] order single pixel approximation is explained based on the simple two pixel model which can be analytically solved. In the calculation of $\rho(r)$, all the symmetry requirements are satisfied and the number of electrons F_{000} is always kept constant through the iteration procedure. The summation concerning $\rho(r)$ in the above equations should be written in the integral form mathematically. Since we must use very limited number of pixels in the numerical calculation, the integral can be replaced by summation in the above mentioned equation.

The evaluation of reflections missing from the summation is possible, after the completion of entropy maximization. In the Fourier summation, the amplitudes of the unobserved reflections are assumed to be zero, but MEM [Collins, 1982] technique provides the most probable values. MEM [Collins, 1982] has many advantages compared to the conventional Fourier method in electron density calculations which are, (i) MEM [Collins, 1982] provides an explicit formulation for the actual electron density rather than the normalized one, (ii) leads to least biased calculation, (iii) performs accurately even with the limited number of information, (iv) unobserved reflections can be simulated, (v) precise electron density maps can be obtained and (vi) the existence of bonding electrons can be clearly visualized.

Owing these advantages over other methods, considerable research on the charge density analysis using the MEM [Collins, 1982] method is going on successfully.

1.11.2.6 MEM methodology

A three dimensional description of the electron density in a selected crystal structure can be evaluated from the X-ray diffraction patterns because X-rays are scattered from the electron clouds of atoms in a crystal lattice. The structure factors retrieved from the Rietveld [Rietveld, 1969] refinement are utilized for the evaluation and visualization of the charge density of the prepared systems.

In this work, the electronic structure and the spatial electron density inside the unit cell have been successfully implemented using sophisticated computer programs. The MEM [Collins, 1982] charge density calculations are performed by adapting the formalism proposed by Collins [1982]. All the data sets are refined using the software PRIMA (PRactice Iterative MEM analysis) [Izumi et al., 2002]. PRIMA is a software for MEM [Collins, 1982] analysis to calculate the electron densities from the X-ray diffraction data. The unit cell was partitioned into $64 \times 64 \times 64$ pixels for cubic systems and the prior electron density in each of the pixel was uniformly fixed as Z/a_0^3, where Z is the total number of electrons in the unit cell and a_0^3 is the volume of the unit cell. The Lagrangian multiplier in each case is selected suitably such that, the convergence criterion C becomes unity after minimum number of iterations. The clear visualization of electron density in 3D and 2D levels was plotted using VESTA (Visualization for Electron and STructural Analysis) [Momma, 2008] software package. VESTA [Momma, 2008] represents the crystal structure by using various models like, ball and stick, space filling, polyhedral and wireframe. With all the advanced features, the software VESTA [Momma, 2008] has been effectively utilized in this work for electronic structure studies.

1.11.3 Optical band gap evaluation using UV-vis spectra

UV-vis absorption analysis is one of the most frequently used methods for optical characterization of material under investigation. The optical band gap can be evaluated from the absorption spectrum using the method proposed by Wood and Tauc [Wood and Tauc, 1972]. The optical band gap is associated with the absorbance (α) and photon energy (hv) by the following relation,

$$\alpha hv = A(hv - E_g)^n \tag{1.43}$$

where α is the absorbance, hv is photon energy, E_g is the optical band gap, A is a constant and n is an index which takes different values for different transition types of transitions, viz., n=1/2, for direct allowed transition, n=3/2 for direct forbidden transition, n=2 for indirect allowed transition and n=3 for indirect forbidden transition. For direct band gap materials, equation (1.43) can be written as,

$$\alpha hv = A(hv - E_g)^{1/2} \tag{1.44}$$

By squaring the equation (1.44),

$$(\alpha hv)^2 = A(hv - E_g) \tag{1.45}$$

$$(\alpha E^2) = Ahv - AE_g \tag{1.46}$$

The above equation resembles the equation of a straight line,

$$y = mx + C \tag{1.47}$$

Comparing equations (1.46) and (1.47),

$$y = (\alpha E^2) \hspace{6cm} (1.48)$$

If y=0, then the equation (1.47) becomes,

$$mx + C = 0 \hspace{6cm} (1.49)$$

$$A(hv - E_g) = 0 \hspace{5cm} (1.50)$$

In the above equation, the constant A cannot be equal to zero, then,

$$hv - E_g = 0 \hspace{6cm} (1.51)$$

$$E_g = hv \hspace{6cm} (1.52)$$

Equation (1.52) gives the optical band gap energy, which can be evaluated using the Tauc's procedure, by fitting a line through the linear portion of the band edge transition. Using UV-vis absorption data, a graph can be drawn by taking energy (hv) in x-axis, and $(\alpha hv)^2$ in y-axis. From this Tauc plot, the extrapolation of the tangent of linear portion of the curve to x-axis, will give the value of the optical band gap E_g.

1.11.4 Grain size evaluation

The average grain size can be evaluated using full width at half maximum (FWHM) of the powder XRD peaks using the Scherrer formula [Cullity and Stock, 2001] which is given by,

$$t = \frac{0.9\lambda}{\beta \cos\theta} \hspace{5cm} (1.53)$$

where, λ is the wavelength of X-ray, which is 1.54056 Å, β is the full width at half maximum (FWHM) in radians, and θ is the Bragg angle, t is the average grain size, which is the average size of coherently diffracting domains.

In this work, the average grain size of all the prepared samples has been evaluated using the GRAIN software [Saravanan, personal communication].

References

[1]   Abramowitz M and Stegun I. A, Handbook of Mathematical Functions, National Bureau of Standards, (1964).

[2]   Agarwal B. K, X-ray spectroscopy, 2nd edition, Springer-verlog, Berlin, (1991). https://doi.org/10.1007/978-3-540-38668-1

[3]   Ali A. I, Kaytbay S. H, Mater Sci Appl. 2, 716 (2011).

[4]   Al-Zoubi A. S, Kishk A. A and Glisson A. W, Prog Electromagn Res. 77, 379 (2007). https://doi.org/10.2528/PIER07082504

[5]   Azaroff L.V, Elements of X-ray crystallography, Mc Graw hill book company, New York, p 79, (1968).

[6]   Badapanda T, Senthil V, Rout S. K, Cavalcante L. S, Simoes A. Z, Sinha T. P, Panigrahi S, de Jesus M. M, Longo E, Varela J. A, Curr. Appl. Phys. 11, 1282 (2011). https://doi.org/10.1016/j.cap.2011.03.056

[7]   Bhaskar Reddy S, Prasad Rao K, Ramachandra Rao M. S, J. Alloy Compd. 481, 692 (2009). https://doi.org/10.1016/j.jallcom.2009.03.075

[8]   Bragg W. L, The diffraction of short electromagnetic waves by a crystal. Proc Cambridge Philoso Soc. 17 (1913).

[9]   Buckner D. A, Wilcox P. D, J Am. Ceramic. Soc. Bull. 51, 218 (1972).

[10]   Buscaglia M. T, Buscaglia V, Viviani M, Nanni P, Hanuskova M, J. Euro. Ceram. Soc. 20, 1997 (2000). https://doi.org/10.1016/S0955-2219(00)00076-5

[11]   Caglioti G, Paoletti A, Ricci F. P, Nucl. Instrum. 3, 223 (1958). https://doi.org/10.1016/0369-643X(58)90029-X

[12]   Carvalho J. F, Hernandes A. C, Cryst. Res. Technol. 40, 847 (2005). https://doi.org/10.1002/crat.200410444

[13]   Cavalcante L. S, Sczancoski J. C, De Vicente F. S, Frabbro M.T, Siu Li. M, Varela J. A, Longo E, J Sol-Gel Sci Technol. 49, 35 (2009). https://doi.org/10.1007/s10971-008-1841-x

[14]   Chen T, Chu S, Juang Y, Sens Actuators B. 102, 11 (2002). https://doi.org/10.1016/S0924-4247(02)00389-8

[15]   Cheng B. L, Wang. C, Wang S. Y, Lu H. B, Zhou Y. L, Chen Z. H, Yang G. Z, J Eur ceram soc. 25, 2295 (2005) https://doi.org/10.1016/j.jeurceramsoc.2005.03.049

[16] Chinor K. U, Ferroelectric Devices, Marcel Dekker INC, New York (2000).

[17] Cho W. S, J. Phys. Chem. Solids. 59, 659 (1998). https://doi.org/10.1016/S0022-3697(97)00227-8

[18] Choi S. H, Ko Y. N, Lee J, Kang Y. C, Ceram Int. 38, 4029 (2012). https://doi.org/10.1016/j.ceramint.2012.01.061

[19] Chonghe L, Kitty C. K. S, Ping W, J. Alloy Compd. 372, 40 (2004). https://doi.org/10.1016/j.jallcom.2003.10.017

[20] Chou, Zhao Z, Zhang W, Zhai J, Mater Des. 31, 3703 (2010). https://doi.org/10.1016/j.matdes.2010.03.006

[21] Collins D. M, Nature. 298, 49 (1982). https://doi.org/10.1038/298049a0

[22] Coppens P, Guru Row T. N, Leung P, Stevens E. D, Becker P. J, Yang Y, Acta Cryst. A35, 63 (1979). https://doi.org/10.1107/S0567739479000127

[23] Coppens P, J. Phys. Chem. 93, 7979 (1989). https://doi.org/10.1021/j100361a006

[24] Cullity B. D, Stock S. R, Elements of X-ray diffraction, Pearson education. 3rd edn. Prentice Hall, Upper Saddle River, 558 (2001).

[25] Debye P, Scherrer P, Phys. Zeit. 19, 474 (1918).

[26] Deshpande S. B, Khollam Y. B, Bhoraskar S. V, Date S. K, Sainkar S. R, Potdar H. S, Mater. Lett. 59, 293 (2005). https://doi.org/10.1016/j.matlet.2004.10.006

[27] Dollase W. A, J Appl Crystallogr. 19, 267 (1986). https://doi.org/10.1107/S0021889886089458

[28] Dou S. X, Ding Y, Eichler H. J, Zhu Y, Ye P.X, Opt. Commun. 131, 322 (1996). https://doi.org/10.1016/0030-4018(96)00275-1

[29] Fan Y, Yu S, Sun R, Li L, Yin Y, Du R, Thinfilms. 518, 3610 (2010).

[30] Funakoshi H, Okamoto A, Sato K, J. Mod. Opt. 52, 1511 (2005). https://doi.org/10.1080/09500340500052929

[31] Goldschmidt V. M, Naturwissenschafsen, 14, 477-485 (1926). https://doi.org/10.1007/BF01507527

[32] Goldstein J, Newbury E, Scanning Electron Microscopy and X-ray Microanalysis, 3rd edition, Plenum Press, New York (2003). https://doi.org/10.1007/978-1-4615-0215-9

[33] Golmohammad M, Nemati Z. A, FaghihiSani M. A, Mater Sci-Poland. 28, 421 (2010).

[34] Gullapalli S, Barron A. Characterization of Group 12-16 (II-VI) Semiconductor Nanoparticles by UV-visible Spectroscopy, OpenStax CNX Web site. http://cnx.org/content/m34601/1.1/, Jun 12, 2010.

[35] Hench L. L and West L. K, Principles of Electronic Ceramics, John Wiley & Sons. Inc, New York p244 (1990).

[36] Hennings D, Schreinemacher B, Schreinemacher H, J. Eur. Ceram. Soc. 81, 13 (1994).

[37] Howard C. J, J Appl Crystallogr. 15, 615 (1982). https://doi.org/10.1107/S0021889882012783

[38] Hwang H, Han Y. H, Jpn. J. Appl. Phys. 701, 39 (2000).

[39] Itoh J, Park D. C, Ohashi N. C, Sakaguchi I, Yashima I, Haneda H, Tanaka J, Jpn. J. Appl. Phys. 41, 3798 (2002). https://doi.org/10.1143/JJAP.41.3798

[40] Izumi F, Dilanien R.A, Recent Research Developments in Physics Part II, Vol.3, Transworld Research Network. Trivandrum, p699–726, (2002).

[41] Jacob R, Harikrishnan G. Nair, Jayakumari, Isac, PAC., 9, 73 (2015). https://doi.org/10.2298/PAC1502073J

[42] Jaffe B, Cook W. R, Jaffe H, Piezoelectric Ceramics, Academic press London and New York, Vol. 3 (1971).

[43] Jain A, Saroh R, Pastor M, Jha A. K, Panwar A.K, Curr. Appl. Phys. 16, 859 (2016). https://doi.org/10.1016/j.cap.2016.04.022

[44] Jana L, Petra S, Trojan M, J. Therm. Anal. Calorim. 93, 823 (2008). https://doi.org/10.1007/s10973-008-9329-z

[45] Jayanthi S, Kutty T. R. N, Mat Sci Eng B. 110, 202 (2004). https://doi.org/10.1016/j.mseb.2004.03.008

[46] Jaynes E. T, IEEE Trans Syst Sci Cybern SSC. 4, 227 (1968). https://doi.org/10.1109/TSSC.1968.300117

[47] Julphunthong P, Chootina S, Bongkarn T, Ceram Int. 39, S415 (2013). https://doi.org/10.1016/j.ceramint.2012.10.105

[48] Jung D. S, Hong S. K, Cho J. S, Kang Y. C, Mater. Res. Bull. 43, 1789 (2008). https://doi.org/10.1016/j.materresbull.2007.07.011

[49] Khanfekr A, Tamizifar M, Naghizadeh R, JNS. 4, 31 (2014).

[50] Klug H. P, Alexander L. E, X-ray diffraction procedures, second edition, John Wiley New York 251 (1959).

[51] Koelzynski A, Smiech K. T, Ferroelectrics, 123, 314 (2005).

[52] Kumar P, Singh P, Singh S, Juneja J. K, Prakash C. and Raina K. K, Ferroelectric Lett. 36, 92 (2009). https://doi.org/10.1080/07315170903152797

[53] Lazarevic Z.Z, Romcevic N.Z, Stojanovic B.D, J Optoelectron Adv M. 10, 2675 (2008).

[54] Li Z. C, Bergman B, Ceram Int. 31, 375 (2005). https://doi.org/10.1016/j.ceramint.2004.06.021

[55] Liu Y, Feng Y, Wu X, Han X, J Alloy Compd. 472, 441 (2009). https://doi.org/10.1016/j.jallcom.2008.04.081

[56] Livingston F. E, Sarney W. L, Niesz K, Ould-Ely T, Tao A.R, Morse D. E, Proc. of SPIE. 7321, 73210I-1 (2009) https://doi.org/10.1117/12.818170

[57] Mancic D, Paunovic V, Vijatovic M, Stojanovic B, Zivkovic L, Sci Sinter. 40, 283 (2008). https://doi.org/10.2298/SOS0803283M

[58] Manoj Kumar, Ashish Garb, Ravi Kumar, Bhatnagar M. C, Physica. B. 403, 1819 (2008). https://doi.org/10.1016/j.physb.2007.10.144

[59] March A. Z Kristallogr. 81, 285 (1932).

[60] Matsuo Y, Sasaki H, J. Am. Ceram. Soc. Bull. 51, 218 (1972).

[61] Melo D. M. A, Cesar A, Martinelli A. E, Silva Z. R, Leite E. R, Longo E, Pizanni P.S, J Solid State Chem. 177, 670 (2004). https://doi.org/10.1016/j.jssc.2003.08.018

[62] Momma K, Izumi F, VESTA: a three-dimensional visualization system for electronic and structural analysis. J. Appl. Crystallogr. 41, 653 (2008). https://doi.org/10.1107/S0021889808012016

[63] Motta F.V, Marques A.P.A, Escote M.T, Melo D.M.A, Ferreira A.G, Longo E, Leite E.R, Varela J.A, J Alloy Comp. 465, 452 (2008). https://doi.org/10.1016/j.jallcom.2007.10.107

[64] Moulson A. J and Herbert J. M, Electroceramics: materials, properties and applications, 2nd edition, Wiley, New York (2003). https://doi.org/10.1002/0470867965

[65] Narayan H, Alemu. H, Macheli L, Rao G, Nanotechnology, 20, 255601 (2009). https://doi.org/10.1088/0957-4484/20/25/255601

[66] Niesz K, Ely T. O, Tsukamoto H, Morse D. E, Ceram Int. 37, 303 (2011). https://doi.org/10.1016/j.ceramint.2010.08.040

[67] Omar A, Abdelal A, Kolthoum I. Othman, Ezzat, Elshazly S, IJSER. 5, Issue 11 (2014).

[68] Park Y, Kim H, J. Am. Ceram. Soc. 80, 106 (1997). https://doi.org/10.1111/j.1151-2916.1997.tb02797.x

[69] Petricek V, Dusek M, Palatinus L, Kristallogr Z, Crystallographic Computing System JANA2006: General features, 229, 345 (2014).

[70] Pradhan S, Roy G. S, Researcher, 5, 63 (2013).

[71] Qian F, Wang X, Wang X, Bu Y, Optik Int. J. Light Electron Opt. 201, 112 (2001).

[72] Rahaman M. N, Ceramic Processing and Sintering, 2nd Edition, Marcel Dekker Inc. (2003).

[73] Rahman S. N, Khatun N, Islam S, Ahmed N. A, IJETCAS. 15, 7 (2014).

[74] Rick U, J. Am. Ceram. Soc. 90, 3326 (2007). https://doi.org/10.1111/j.1551-2916.2007.01881.x

[75] Rehrig P. W, Messing G. L and Trolier-McKinstry S, J. Am. Ceram. Soc. 83, 2654 (2000). https://doi.org/10.1111/j.1151-2916.2000.tb01610.x

[76] Rietveld H. M, Acta Crystallogr. 22, 151 (1967). https://doi.org/10.1107/S0365110X67000234

[77] Rietveld H. M, J. Appl. Crystallogr. 2, 65 (1969). https://doi.org/10.1107/S0021889869006558

[78] Rout S.K, Cavalcante L.S, Sczancoski J.C, Badapanda T, Panigrahi S, SiuLi M, Longo E, Physica. B. 404, 3341 (2009). https://doi.org/10.1016/j.physb.2009.05.014

[79] Sahoo P. S, Panigrahi A, Patri S. K, Choudhary R. N. P, J. Alloys Compd. 484, 832 (2009). https://doi.org/10.1016/j.jallcom.2009.05.051

[80] Sakata M, Sato M, Acta Cryst. A46, 263 (1990). https://doi.org/10.1107/S0108767389012377

[81] Sanna S, Thierfelder C, Wippermann S, Sinha T. P and Schmidt W. G, Phys. Rev B. 83, 054112 (2011). https://doi.org/10.1103/PhysRevB.83.054112

[82] Sarangi S, Badapanda T, Behera B, Anwar S, Rehrig P.W, Park S.E, Kinstry T.M.S, Messing G.L, Jones B, Shrout T.R, J. Appl. Phys. 86, 1657 (1999). https://doi.org/10.1063/1.370943

[83] Saravanan R, GRAIN software, Personal communication.

[84] Saravanan R, Rani M. P, Metal and Alloy Bonding- an Experimental Analysis, Chapter 2 Springer-Verlag London Press, (2012). https://doi.org/10.1007/978-1-4471-2204-3

[85] Schwartz R. W, Chem. Mater. 9, 2327 (1997).

[86] Seeharaj P. N, Boonchom B, Charoonsuk P, Lohsoontornd P. K, Vittayakorna N, Ceram. Int. 39, S559 (2013). https://doi.org/10.1016/j.ceramint.2012.10.135

[87] Shieh J, Yeh J. H, Shu Y. C, Yen J. H, Mater Sci Eng B. 161, 50 (2015). https://doi.org/10.1016/j.mseb.2008.11.046

[88] Skoog D. A, Holler F. J, Crouch S. R, Principles of Instrumental Analysis. Sixth Edition, Thomson Brooks, USA, (2007).

[89] Smaalen S.V, Jeanette Netzel J, Phys. Scr. 79, 048304 (2009). https://doi.org/10.1088/0031-8949/79/04/048304

[90] Stout G. H, Jensen L. H, X-ray structure determination-a practical guide. The Macmillan Company Collier-Macmillan, London, 217 (1968).

[91] Takata M, Nishibori E, Shinmura M, Tanaka H, Tanigaki K, Kosaka M, Sakata M Mater Sci Eng. A312, 66 (2001). https://doi.org/10.1016/S0921-5093(00)01894-3

[92] Takata M, Umeda B, Nishibori E, Sakata M, Saito Y, Ohno M, Shinohara H, Nature, 377, 46 (1995). https://doi.org/10.1038/377046a0

[93] Tang P, Towner D. J, Hamano T, Meier A. L, Wessels B, Optics Express, 12, 5962 (2004). https://doi.org/10.1364/OPEX.12.005962

[94] Thompson P, Cox D. E, Hastings J. B, J Appl Crystallogr. 20, 79 (1987). https://doi.org/10.1107/S0021889887087090

[95] Vijatovic M. M, Bobic J. D, Stojanovic B. D, Sci Sinter. 40, 155 (2008). https://doi.org/10.2298/SOS0802155V

[96] Wang Z, Jiang S, Li G, Xi M, Li T, Ceram. Int. 33, 1105 (2007). https://doi.org/10.1016/j.ceramint.2006.03.015

[97]   Warren B. E, X-ray diffraction, Chapter 3. Dover publications, New York (1990).

[98]   Waugh, Mark D, Electronic Engineering Times, (2010).

[99]   Wertheim G. K, Butler M. A, West K. W, Buchanan D. N. E, Rev Sci Instrum. 45,
       1369 (1974). https://doi.org/10.1063/1.1686503

[100] Wood D. L, Tauc J, Phys Rev B. 5, 3144 (1972).
       https://doi.org/10.1103/PhysRevB.5.3144

[101] Wyckoff R. W. G, Crystal Structures. Inter-Space Publishers, London, Vol.2,
       p.401, (1963).

[102] Xu Q, Zhang X. F, Huang Y. H, Chen W, Liu H. X, Chen M, Kim B. H, J. Alloys
       Compd. 488, 448 (2009). https://doi.org/10.1016/j.jallcom.2009.08.153

[103] Yoo H. L, Lee S. W, Lee C. E, J. Electroceram. 10, 215 (2003).
       https://doi.org/10.1023/B:JECR.0000011220.03915.a4

[104] Yoon S. H, Lee K. H, Kim H, J. Am. Ceram. Soc. 83, 2463 (2000).
       https://doi.org/10.1111/j.1151-2916.2000.tb01577.x

[105] Zhang Q, Zhai J, Kong L, Yao X, J. Appl. Phys. 112, 124112 (2012).
       https://doi.org/10.1063/1.4771669

[106] Zhou Z. G, Tang Z. L, Zhang Z. T, Wlodarski W, Sens Actuators B. 77, 22 (2001).
       https://doi.org/10.1016/S0925-4005(01)00667-0

Chapter 2

Results

Abstract

Chapter 2 presents the results obtained from various experimental characterization and analytical techniques performed to investigate the five differently doped $BaTiO_3$ ceramic systems. The plots of experimental X-ray diffraction patterns, Rietveld fitted profiles, UV-visible absorption graphs, Tauc plots, SEM micrographs, EDS spectra, 3-dimensional (3D), 2-dimensional (2D) electron density contour maps and 1-dimensional (1D) electron density line profiles are presented in this chapter. The tables of structural parameters refined from the Rietveld method and optical band gap values with respect to the dopant concentration are also given. The elemental compositions of the prepared ceramics from EDS analysis, bond lengths and mid-bond electron density values from MEM analysis are also presented.

Keywords

Results, Barium Titanate, Doped $BaTiO_3$, Powder XRD, EDS, Charge Density, SEM, Optical

Contents

2.1 Introduction

In this work, five differently doped $BaTiO_3$ perovskite structured ceramic systems have been synthesized by solid state reaction technique. The influence of various dopants such as, strontium (Sr), zirconium (Zr), lanthanum (La), cerium (Ce) and strontium and zirconium (Sr & Zr, (co-doping)) in the crystal structure of $BaTiO_3$ ceramic systems has been thoroughly investigated using various characterization techniques. The results obtained from various characterization techniques and analytical methods are reported in this chapter.

The structural characterization has been carried out using powder X-ray diffraction (PXRD). The experimental XRD data have been subjected to powder profile refinement using the Rietveld method [Rietveld, 1969] through JANA 2006 [Petricek et al., 2014] software. The experimental XRD patterns, Rietveld [Rietveld, 1969] fitted profiles, structural parameters such as cell constant, unit cell volume and density etc., determined from the Rietveld [Rietveld, 1969] refinement are given in section 2.2.

The optical characterization of the prepared samples has been done using UV-visible absorption spectroscopy (UV-vis). From the UV-vis absorption data, optical band gap value has been evaluated using the procedure given by Wood and Tauc [Wood and Tauc, 1972]. The optical band gap energy values for all the doped titanates studied in this work evaluated from Tauc plot methodology [Wood and Tauc, 1972] are presented in section 2.3.

The surface morphology and microstructure of all the doped $BaTiO_3$ ceramic systems have been analyzed using scanning electron microscopy (SEM) by recording SEM images with various magnifications. The SEM micrographs for the five different doped series of solid solutions have been given in section 2.4. Elemental confirmation of the doped $BaTiO_3$ samples has been carried out using energy dispersive X-ray spectroscopy (EDS). The EDS spectra and the atomic and mass percentages of the doped samples are given in section 2.5.

The precise electronic structure, inter-atomic bonding and electron density distribution of the doped samples have been examined through the high resolution maximum entropy method (MEM) [Collins, 1982]. The structure factors extracted from the Rietveld refinement [Rietveld, 1969] have been utilized for MEM [Collins, 1982] calculations. The MEM [Collins, 1982] refinement has been carried out using the software PRIMA [Izumi, 2002]. The three dimensional, two dimensional and one dimensional electron density distributions have been plotted and analyzed using the visualization software VESTA [Momma, 2008]. The 3D, 2D electron density distributions, 1D line profiles, bond length and electron density values are presented in section 2.6.

2.2   Structural characterization – Powder X-ray diffraction

2.2.1 $Ba_{1-x}Sr_xTiO_3$

$Ba_{1-x}Sr_xTiO_3$ solid solutions with various Sr concentrations (x=0.2, 0.4 & 0.6) were synthesized by the solid state reaction technique and characterized by powder X-ray diffraction (PXRD). The XRD patterns were collected in the 2θ range of $10°$-$120°$ with the step size of $0.02°$.

The experimental X-ray diffraction patterns of $Ba_{1-x}Sr_xTiO_3$ (x=0.2, 0.4 & 0.6) are shown in figure 2.1(a). Figure 2.1(b) shows the enlarged Bragg peaks corresponding to (210) and (211) planes. Figures 2.2(a), (b) & (c) indicate the Rietveld [Rietveld, 1969] fitted profiles of $Ba_{1-x}Sr_xTiO_3$ for various Sr concentrations, x=0.2, 0.4 & 0.6 respectively. In these figures, the cross marks show the observed profiles, continuous lines show the calculated profiles and the vertical lines show the positions of Bragg peaks. The structural parameters refined from the Rietveld method [Rietveld, 1969] are given in table 2.1.

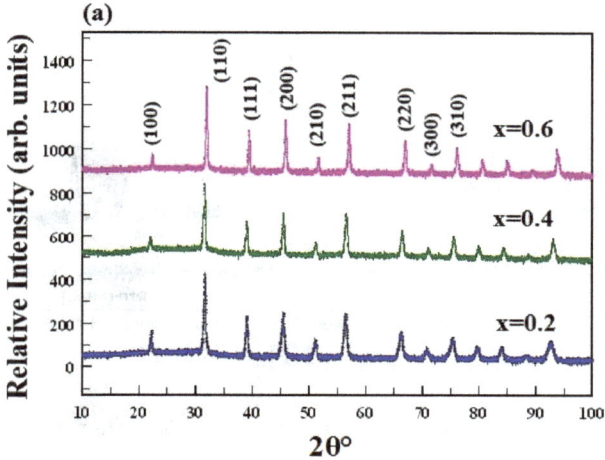

Figure 2.1(a) *Observed powder XRD patterns of $Ba_{1-x}Sr_xTiO_3$ (x=0.2, 0.4 & 0.6).*

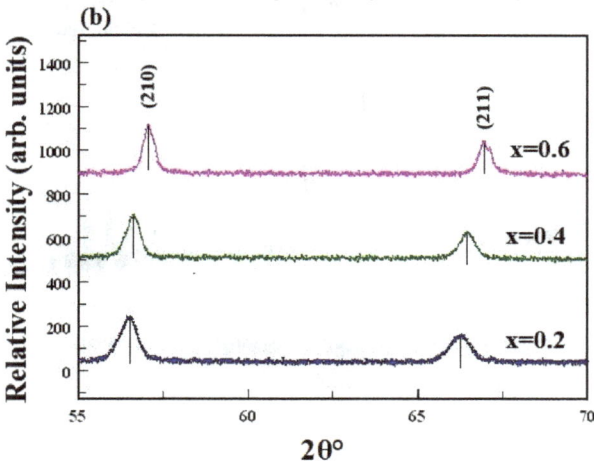

Figure 2.1(b) *Enlarged Bragg peaks of (210) & (211) planes.*

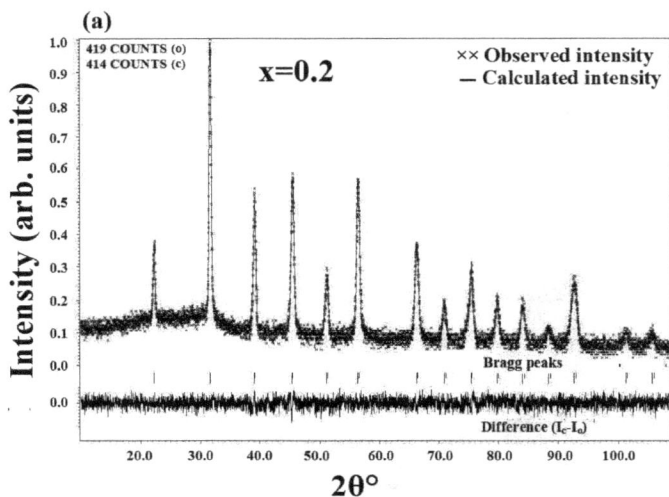

Figure 2.2(a) Fitted XRD profile of $Ba_{1-x}Sr_xTiO_3$, $x=0.2$.

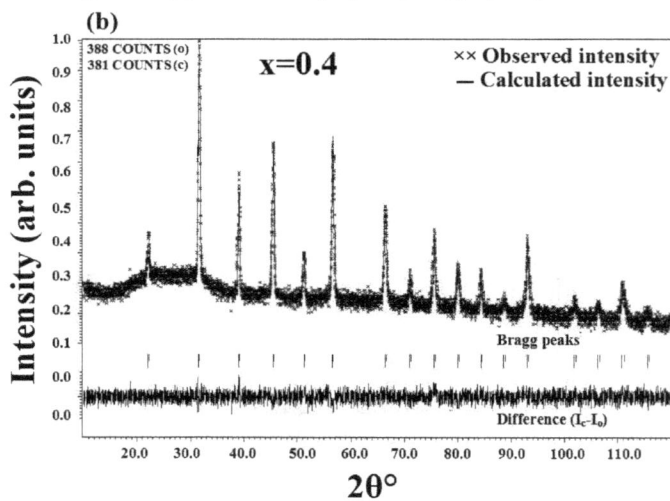

Figure 2.2(b) Fitted XRD profile of $Ba_{1-x}Sr_xTiO_3$, $x=0.4$.

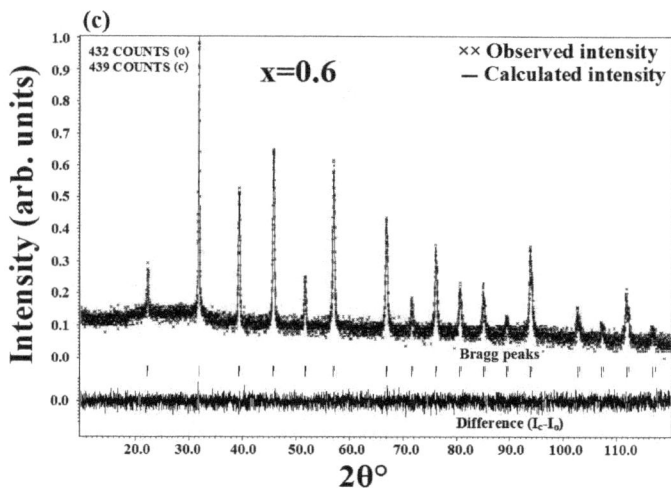

(c)

432 COUNTS (o)
439 COUNTS (c)

x=0.6

×× Observed intensity
— Calculated intensity

Bragg peaks

Difference (I_c-I_o)

$2\theta°$

Figure 2.2(c) *Fitted XRD profile of $Ba_{1-x}Sr_xTiO_3$, x=0.6.*

Table 2.1 *Structural parameters of $Ba_{1-x}Sr_xTiO_3$ refined from Rietveld method.*

Parameters	x=0.2	x=0.4	x=0.6
a=b=c (Å)	3.970(14)	3.962(6)	3.941(4)
$\alpha =\beta =\gamma$ (°)	90	90	90
Unit cell Volume ($Å^3$)	62.60(2)	62.22(9)	61.23(6)
Density (gm/cc)	5.920	5.691	5.510
R_P (%)	9.81	7.21	9.81
R_{obs}(%)	2.81	3.89	3.16
GOF	0.95	0.78	0.90
$F_{(000)}$	102	103	103

R_P - Reliability index for profile

R_{obs} - Reliability index for observed structure factors

GOF - Goodness of fit

$F_{(000)}$ - Number of electrons in the unit cell

2.2.2 BaTi$_{1-x}$Zr$_x$O$_3$

BaTi$_{1-x}$Zr$_x$O$_3$ ceramic systems with various Zr concentrations, x=0.00, 0.04 & 0.06 were prepared by high temperature solid state reaction technique and characterized by powder X-ray diffraction (PXRD) for structural analysis. The XRD patterns are collected in the 2θ range of 10°-120° with the step size of 0.02°. The observed X-ray diffraction patterns of BaTi$_{1-x}$Zr$_x$O$_3$ (x=0.00, 0.04 & 0.06) are given in figure 2.3(a). Figure 2.3(b) shows the enlarged Bragg peaks corresponding to (110) and (111) planes. Figures 2.4(a), (b) & (c) show the Rietveld [Rietveld, 1969] fitted profiles of BaTi$_{1-x}$Zr$_x$O$_3$ with various Zr concentrations, x=0.00, 0.04 & 0.06 respectively. Table 2.2 gives the structural parameters refined from the Rietveld method [Rietveld, 1969].

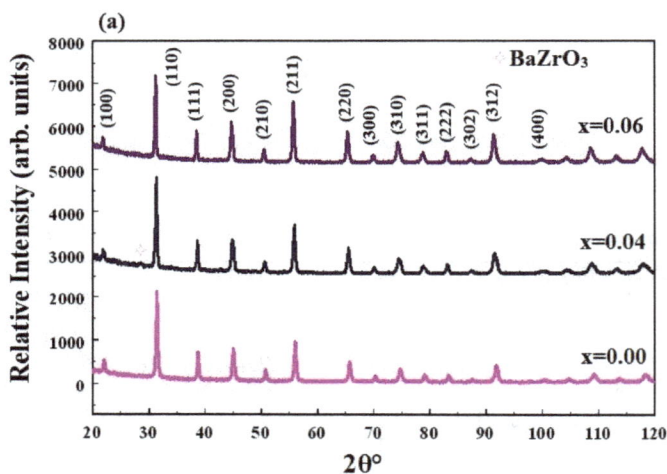

Figure 2.3(a) *Observed powder XRD patterns of BaTi$_{1-x}$Zr$_x$O$_3$ (x=0.00, 0.04 & 0.06).*

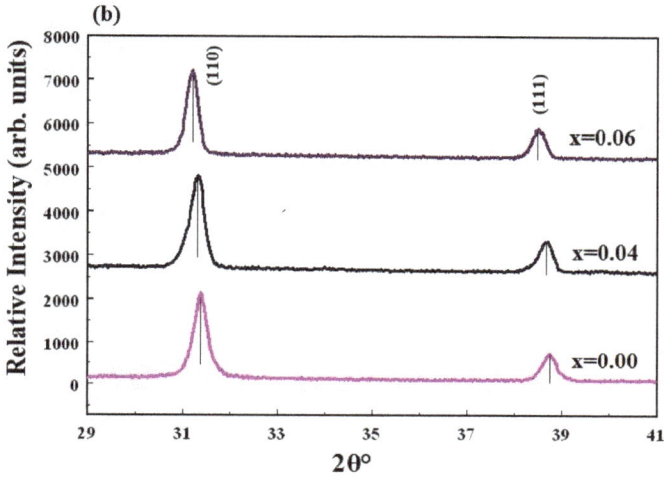

Figure 2.3(b) Enlarged Bragg peaks of (110) & (111) planes.

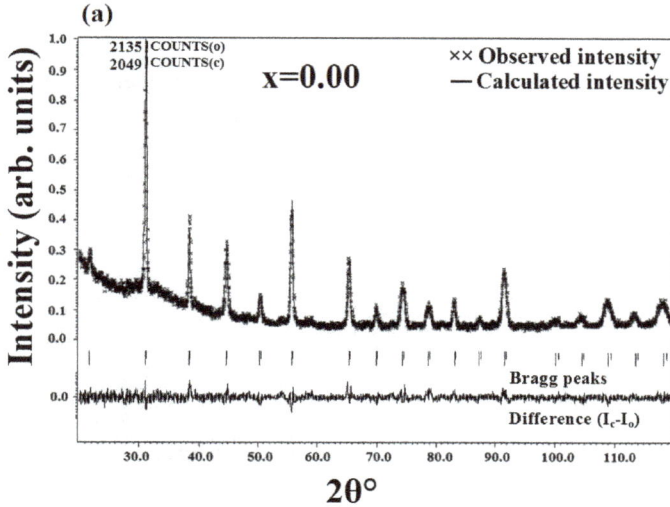

Figure 2.4(a) Fitted XRD profile of $BaTi_{1-x}Zr_xO_3$, x=0.00.

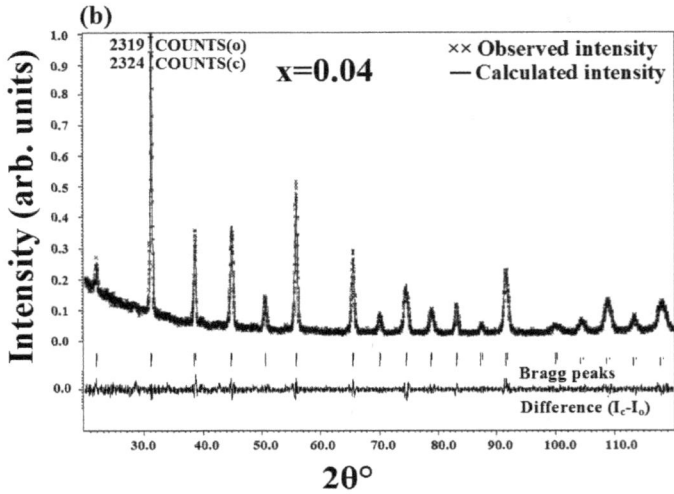

Figure 2.4(b) *Fitted XRD profile of BaTi$_{1-x}$Zr$_x$O$_3$, x=0.04.*

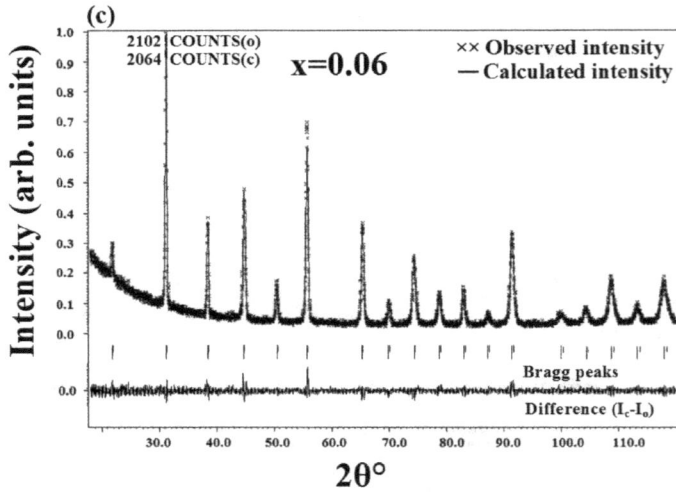

Figure 2.4(c) *Fitted XRD profile of BaTi$_{1-x}$Zr$_x$O$_3$, x=0.06.*

Table 2.2 *Structural parameters of* $BaTi_{1-x}Zr_xO_3$ *refined from the Rietveld method.*

Parameters	x=0.00	x=0.04	x=0.06
a=b=c (Å)	4.006 (9)	4.017 (5)	4.021(4)
α=β=γ (°)	90	90	90
Unit cell volume (Å³)	64.32(5)	64.86(1)	65.02(3)
Density (gm/cc)	6.014(1)	6.017 (2)	6.020 (2)
R_P(%)	6.31	7.38	6.43
R_{obs} (%)	2.05	2.68	1.31
GOF	1.06	1.22	1.11
$F_{(000)}$	102	103	103

R_P - Reliability index for profile

R_{obs} - Reliability index for observed structure factors

GOF - Goodness of fit

$F_{(000)}$ - Number of electrons in the unit cell

2.2.3 $Ba_{1-x}La_{2x/3}TiO_3$

$Ba_{1-x}La_{2x/3}TiO_3$ solid solutions with various La concentrations, x=0.000, 0.005, 0.015, 0.020 & 0.025 were prepared by the chemical method and characterized by powder X-ray diffraction (PXRD). The XRD patterns were collected in the 2θ range of 10°-120° with the step size of 0.02°. The observed XRD patterns of $Ba_{1-x}La_{2x/3}TiO_3$ (x=0.000, 0.005, 0.015, 0.020 & 0.025) are given in figure 2.5(a). Figure 2.5(b) shows the enlarged Bragg peaks corresponding to (101) and (111) planes. Figures 2.6(a), (b), (c), (d) & (e) show the Rietveld [Rietveld, 1969] fitted profiles of $Ba_{1-x}La_{2x/3}TiO_3$ with various La concentrations, x=0.000, 0.005, 0.015, 0.020 & 0.025 respectively. The structural parameters evaluated from the Rietveld method [Rietveld, 1969] are given in table 2.3.

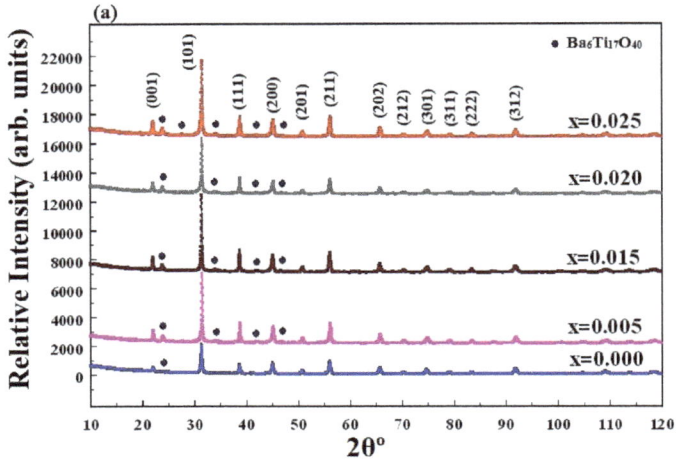

Figure 2.5(a) *Observed powder XRD patterns of $Ba_{1-x}La_{2x/3}TiO_3$ (x=0.000, 0.005, 0.015, 0.020 & 0.025).*

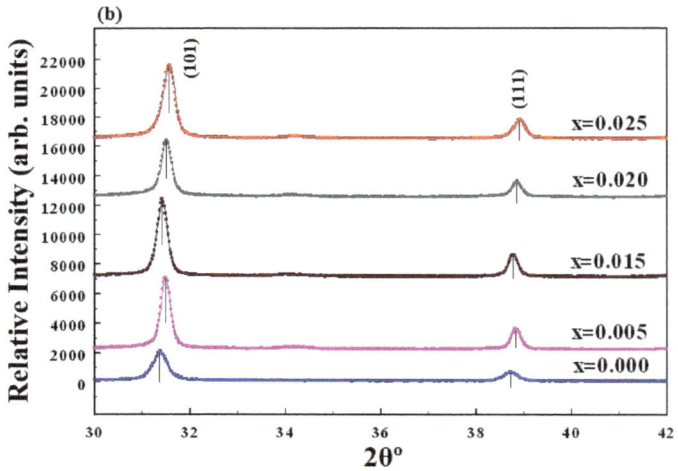

Figure 2.5(b) *Enlarged Bragg peaks of (101) & (111) planes.*

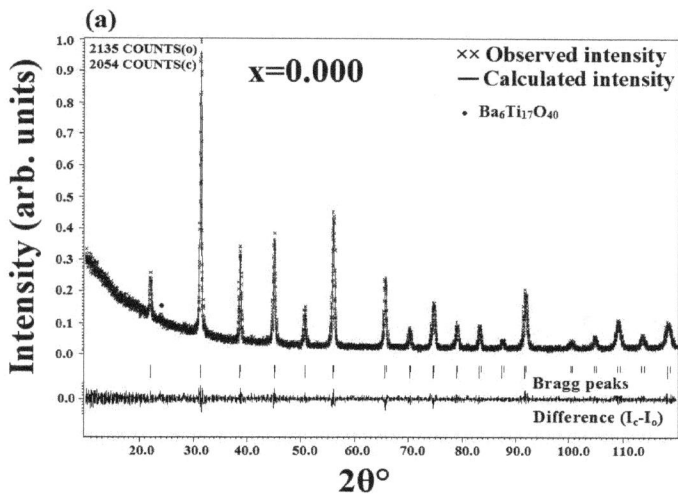

Figure 2.6(a) *Fitted XRD profile of $Ba_{1-x}La_{2x/3}TiO_3$, x=0.000.*

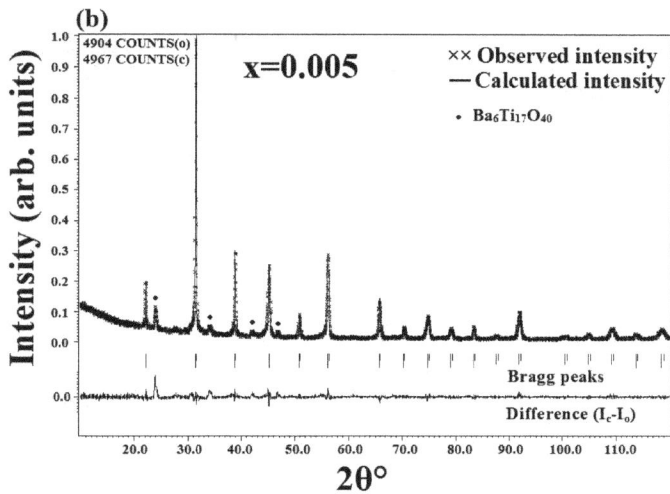

Figure 2.6(b) *Fitted XRD profile of $Ba_{1-x}La_{2x/3}TiO_3$, x=0.005.*

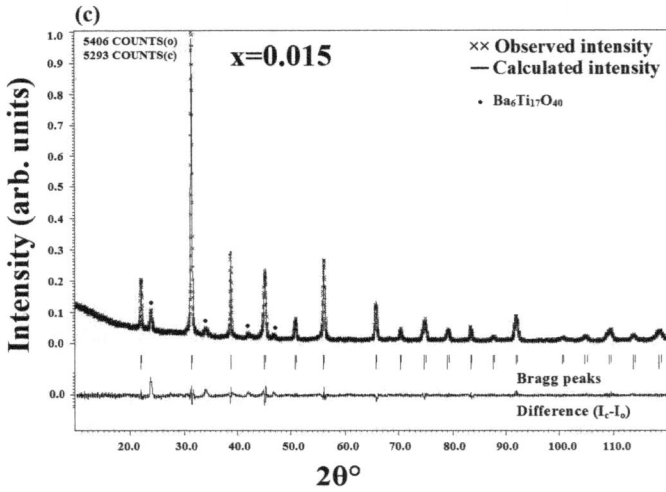

Figure 2.6(c) *Fitted XRD profile of* $Ba_{1-x}La_{2x/3}TiO_3$, $x=0.015$.

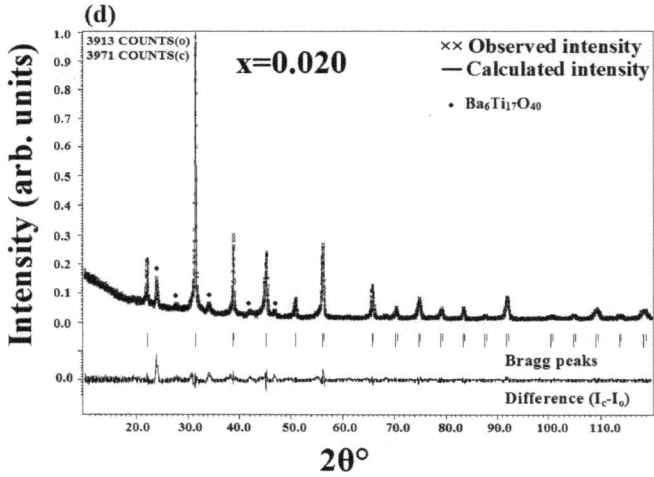

Figure 2.6(d) *Fitted XRD profile of* $Ba_{1-x}La_{2x/3}TiO_3$, $x=0.020$.

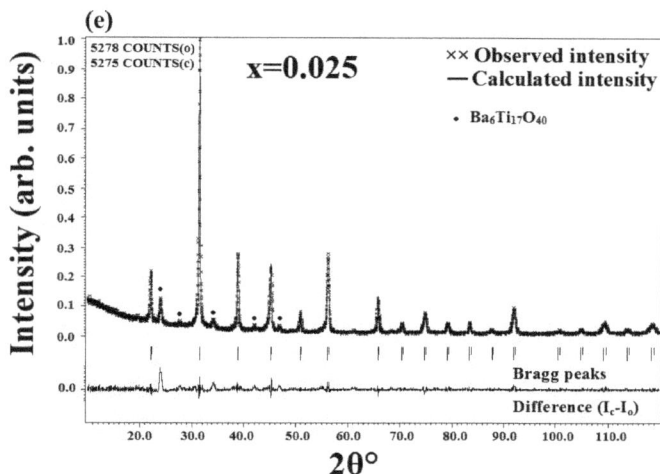

Figure 2.6(e) *Fitted XRD profile of $Ba_{1-x}La_{2x/3}TiO_3$, $x=0.025$.*

Table 2.3 *Structural parameters of $Ba_{1-x}La_{2x/3}TiO_3$ refined from Rietveld method.*

Parameters	x=0.000	x=0.005	x=0.015	x=0.020	x=0.025
a=b (Å)	4.002(11)	4.000(6)	3.999(3)	3.995(9)	3.994(10)
c(Å)	4.017(14)	4.022(8)	4.021(2)	4.017(11)	4.016(2)
$\alpha = \beta = \gamma$ (°)	90	90	90	90	90
Unit cell volume (Å³)	64.35(32)	64.37(24)	64.32(12)	63.12(9)	63.08(17)
Density (gm/cc)	6.016	6.000	6.001	6.013	6.009
R_P (%)	6.38	7.99	7.75	8.40	7.88
R_{obs} (%)	1.98	2.28	2.32	2.77	2.27
GOF	1.06	1.50	1.49	1.53	1.58
$F_{(000)}$	102	102	102	102	102

R_P - Reliability index for profile

R_{obs} - Reliability index for observed structure factors

GOF- Goodness of fit

$F_{(000)}$- Number of electrons in the unit cell

2.2.4 BaTi$_{1-x}$Ce$_x$O$_3$

BaTi$_{1-x}$Ce$_x$O$_3$ ceramic compounds with various Ce concentrations, x=0.02, 0.04, 0.06 & 0.08 were synthesized by the solid state reaction technique and characterized by powder X-ray diffraction (PXRD). The XRD patterns were collected in the 2θ range of 10°-120° with the step size of 0.02°. The observed XRD patterns of BaTi$_{1-x}$Ce$_x$O$_3$ (x=0.02, 0.04, 0.06 & 0.08) are given in the figure 2.7(a). Figure 2.7(b) shows the enlarged Bragg peaks correspond to (220) plane. Figures 2.8(a), (b), (c) & (d) show the Rietveld [Rietveld, 1969] fitted profiles of BaTi$_{1-x}$Ce$_x$O$_3$ with various Ce concentrations x=0.02, 0.04, 0.06 & 0.08 respectively. Structural parameters refined from the Rietveld method [Rietveld, 1969] are given in table 2.4.

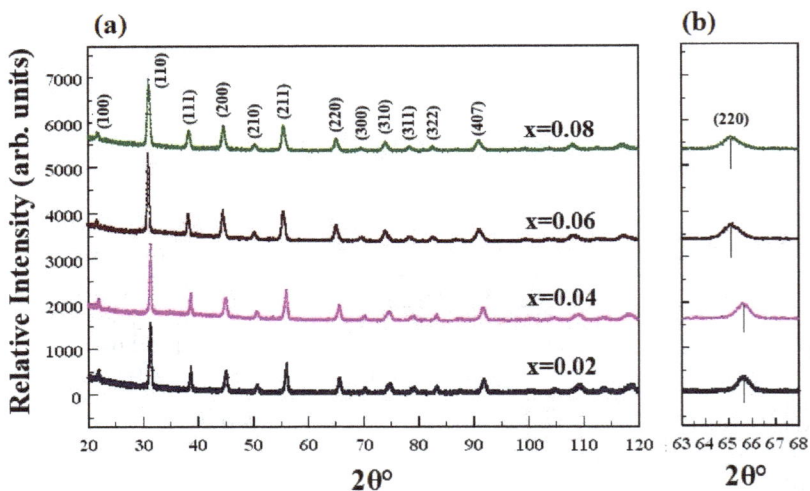

Figure 2.7(a) *Observed powder XRD patterns of BaTi$_{1-x}$Ce$_x$O$_3$ (x=0.02, 0.04, 0.06 & 0.08),* **(b)** *Enlarged Bragg peaks of (220) plane.*

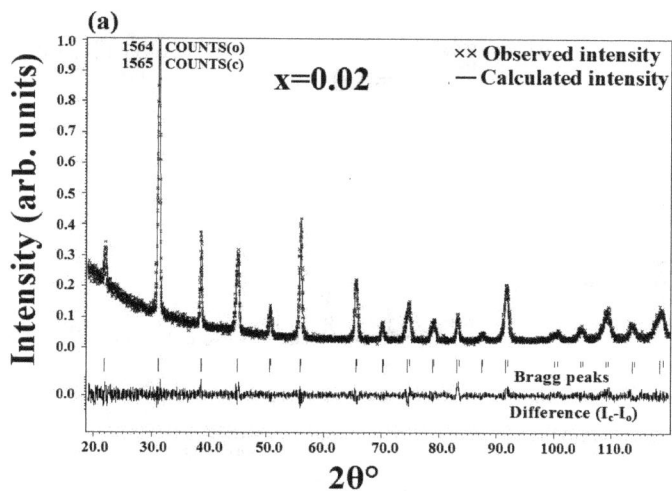

Figure 2.8(a) Fitted XRD profile of $BaTi_{1-x}Ce_xO_3$, $x=0.02$.

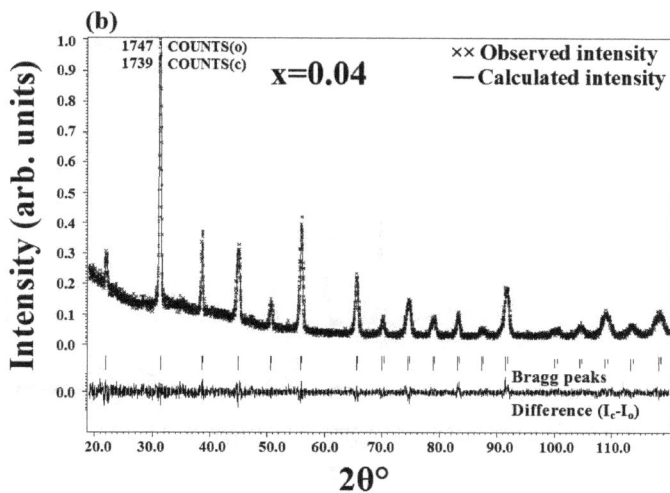

Figure 2.8(b) Fitted XRD profile of $BaTi_{1-x}Ce_xO_3$, $x=0.04$.

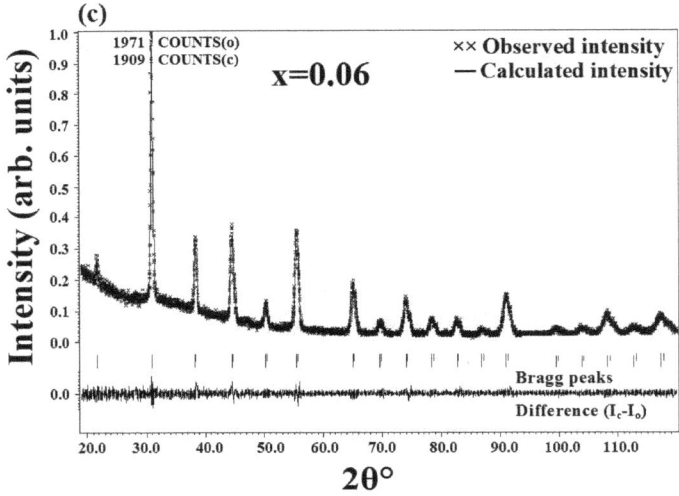

Figure 2.8(c) Fitted XRD profile of $BaTi_{1-x}Ce_xO_3$, x=0.06.

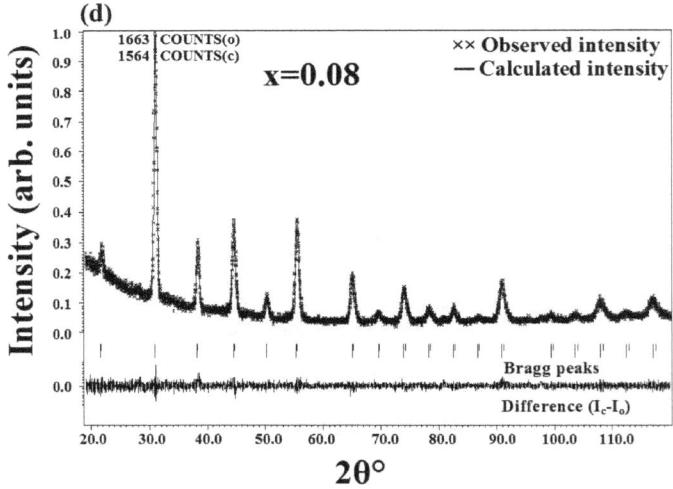

Figure 2.8(d) Fitted XRD profile of $BaTi_{1-x}Ce_xO_3$, x=0.08.

Table 2.4 *Structural parameters of BaTi$_{1-x}$Ce$_x$O$_3$ refined from Rietveld method.*

Parameters	x=0.02	x=0.04	x=0.06	x=0.08
a=b=c (Å)	4.008(8)	4.015(11)	4.030(16)	4.042(18)
α=β=γ (°)	90	90	90	90
Unit cell volume (Å^3)	64.42(1)	64.72(1)	65.52(2)	66.07(3)
Density (gm/cc)	6.056(2)	6.075(1)	6.001(2)	6.044(2)
R$_P$ (%)	7.94	6.99	6.87	6.68
R$_{obs}$ (%)	2.32	2.32	2.55	2.82
GOF	1.15	1.12	1.15	1.06
F$_{(000)}$	103	103	103	105

R$_P$ - Reliability index for profile

R$_{obs}$ -Reliability index for observed structure factors

GOF- Goodness of fit

F$_{(000)}$ - Number of electrons in the unit cell

2.2.5 Ba$_{1-x}$Sr$_x$Ti$_{0.9}$Zr$_{0.1}$O$_3$

Ba$_{1-x}$Sr$_x$Ti$_{0.9}$Zr$_{0.1}$O$_3$ ceramic solid solutions with various Sr concentrations, x=0.00, 0.05, 0.07 & 0.14 were prepared by the solid state reaction technique and characterized by powder X-ray diffraction (PXRD). The XRD patterns were collected in the 2θ range of 10°-120° with the step size of 0.02°. Figure 2.9(a) shows the experimental X-ray diffraction patterns of Ba$_{1-x}$Sr$_x$Ti$_{0.9}$Zr$_{0.1}$O$_3$ (x=0.00, 0.05, 0.07 & 0.14). Figure 2.9(b) shows the enlarged Bragg peaks corresponding to (220) plane. Figures 2.10(a), (b), (c) & (d) show the Rietveld [Rietveld, 1969] fitted profiles of Ba$_{1-x}$Sr$_x$Ti$_{0.9}$Zr$_{0.1}$O$_3$ with various Sr concentrations x=0.00, 0.05, 0.07 & 0.14 respectively. Table 2.5 gives the structural parameters refined from the Rietveld method [Rietveld, 1969].

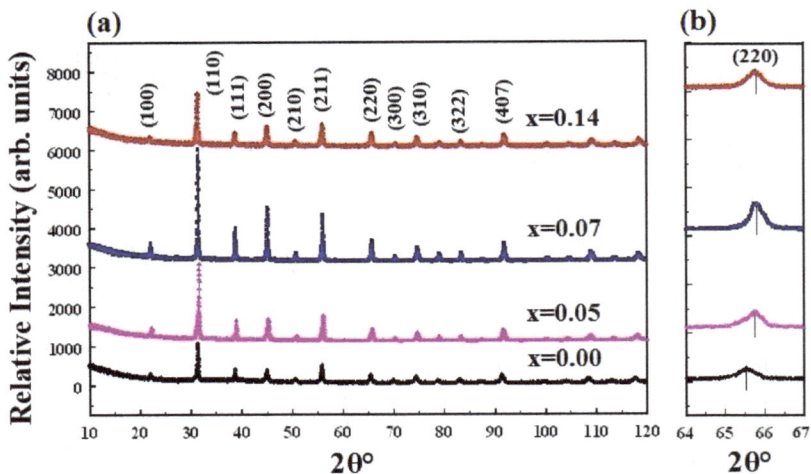

Figure 2.9(a) *Observed powder XRD patterns of* $Ba_{1-x}Sr_xTi_{0.9}Zr_{0.1}O_3$ *(x=0.00, 0.05, 0.07 & 0.14),* **(b)** *Enlarged Bragg peaks of (220) plane.*

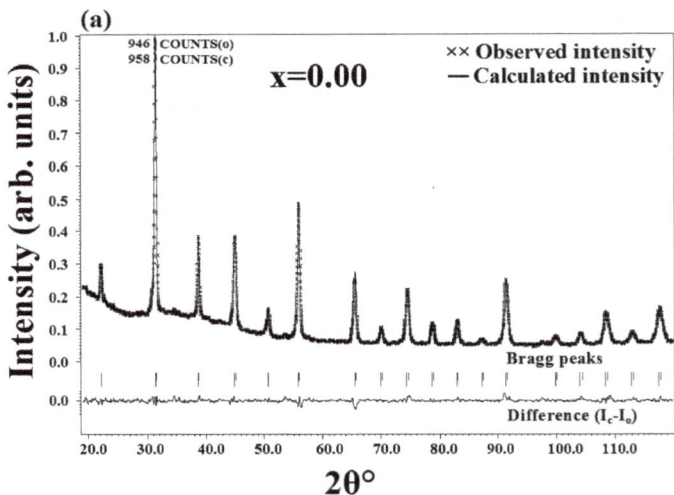

Figure 2.10(a) *Fitted XRD profile of* $Ba_{1-x}Sr_xTi_{0.9}Zr_{0.1}O_3$, *x=0.00.*

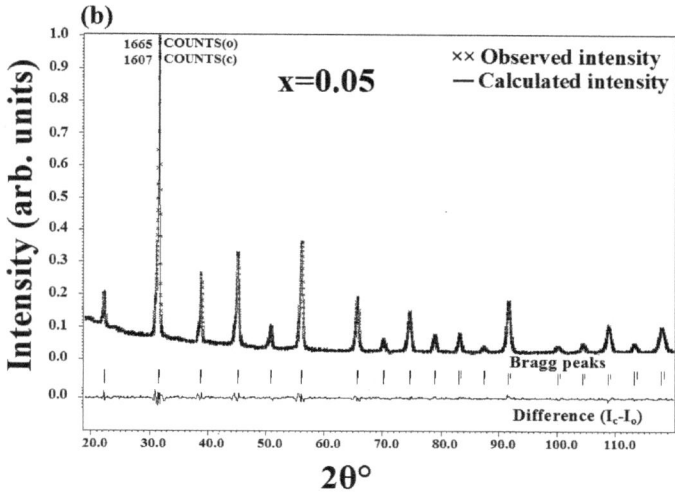

Figure 2.10(b) *Fitted XRD profile of $Ba_{1-x}Sr_xTi_{0.9}Zr_{0.1}O_3$, $x=0.05$.*

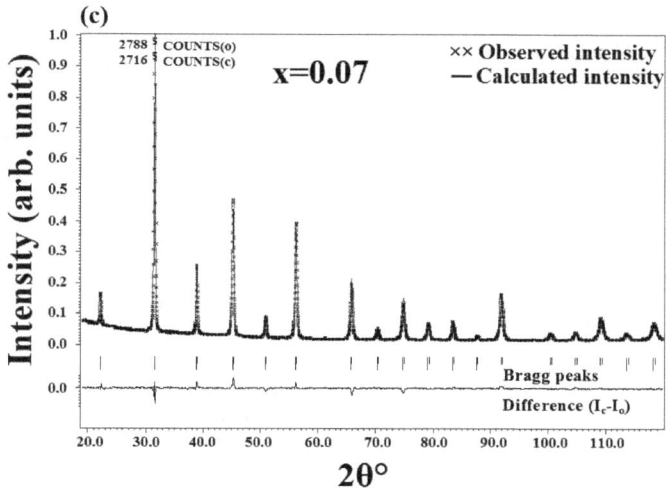

Figure 2.10(c) *Fitted XRD profile of $Ba_{1-x}Sr_xTi_{0.9}Zr_{0.1}O_3$, $x=0.07$.*

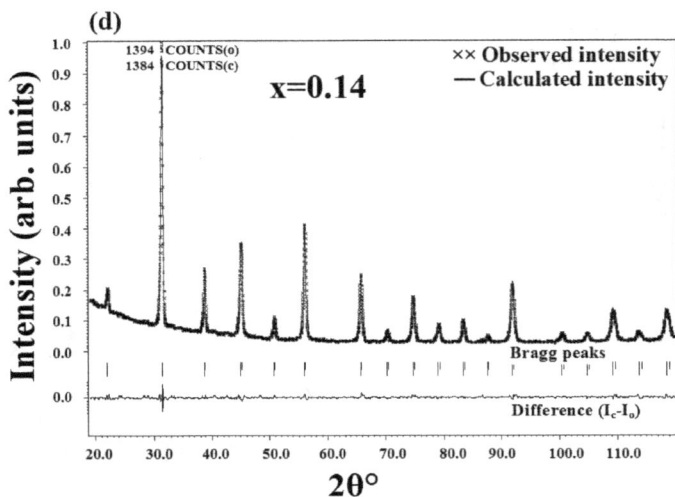

(d)

Figure 2.10(d) Fitted XRD profile of $Ba_{1-x}Sr_xTi_{0.9}Zr_{0.1}O_3$, $x=0.14$.

Table 2.5 Structural parameters of $Ba_{1-x}Sr_xTi_{0.9}Zr_{0.1}O_3$ refined from the Rietveld method.

Parameters	x=0.00	x=0.05	x=0.07	x=0.14
a=b=c (Å)	4.031(8)	4.025(9)	4.012(2)	4.005(6)
α=β=γ (°)	90	90	90	90
Unit cell volume (Å³)	65.43(1)	65.22(1)	64.57(2)	64.12(3)
Density (gm/cc)	6.02(2)	5.98(1)	6.02(2)	5.96(2)
R_P (%)	3.39	8.32	7.14	8.09
R_{obs} (%)	2.53	2.14	2.99	4.17
GOF	0.48	1.15	1.10	1.11
$F_{(000)}$	104	103	103	101

R_P - Reliability index for profile

R_{obs} - Reliability index for observed structure factors

GOF- Goodness of fit

$F_{(000)}$ - Number of electrons in the unit cell

2.3 Optical characterization – UV -visible absorption spectra

2.3.1 $Ba_{1-x} Sr_x TiO_3$

The optical band gap value has been evaluated using UV-vis absorption data obtained in the wavelength range of 200 nm to 2000 nm. Figure 2.11 shows the UV-visible absorption spectra of $Ba_{1-x}Sr_xTiO_3$. The optical band gap was evaluated using the method proposed by Wood and Tauc [Wood and Tauc, 1972]. The Tauc plot of $Ba_{1-x}Sr_xTiO_3$ ceramic system has been drawn using the Tauc's relation [Wood and Tauc, 1972] and is given in figure 2.12. The optical band gap values for $Ba_{1-x}Sr_xTiO_3$ with three different Sr concentrations, x=0.2, 0.4 & 0.6 are given in table 2.6.

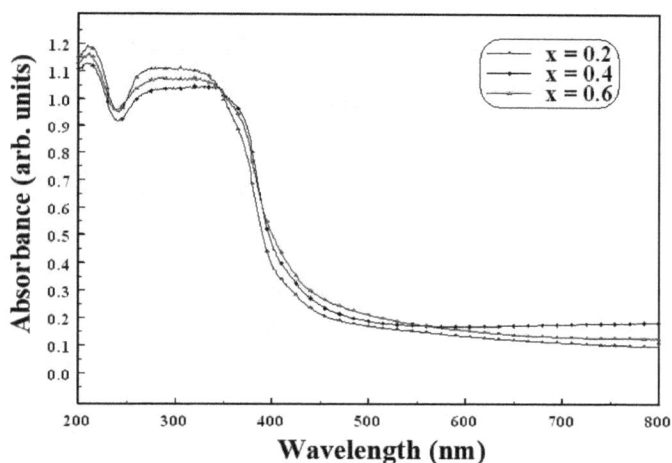

Figure 2.11 UV-visible absorption spectra of $Ba_{1-x}Sr_xTiO_3$.

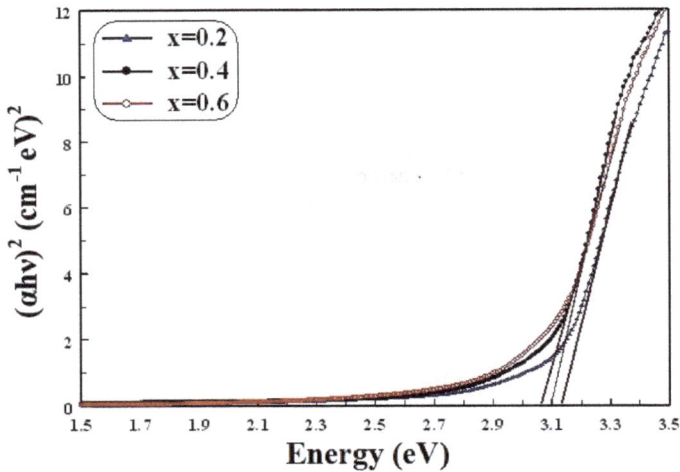

Figure 2.12 *Tauc plot of Ba$_{1-x}$Sr$_x$TiO$_3$.*

Table 2.6 *Optical band gap values of Ba$_{1-x}$Sr$_x$TiO$_3$.*

Samples	Band gap (eV)
x=0.2	3.128
x=0.4	3.098
x=0.6	3.067

2.3.2 BaTi$_{1-x}$Zr$_x$O$_3$

The optical band gap value has been evaluated using UV-vis data obtained in the wavelength range of 200 nm to 2000 nm. Figure 2.13 shows the UV-visible absorption spectra of BaTi$_{1-x}$Zr$_x$O$_3$. The Tauc plot of BaTi$_{1-x}$Zr$_x$O$_3$ has been drawn using the Tauc's relation [Wood and Tauc, 1972] and is given in figure 2.14. The optical band gap values for BaTi$_{1-x}$Zr$_x$O$_3$ with three different Zr concentrations, x=0.00, 0.04 & 0.06 are presented in table 2.7.

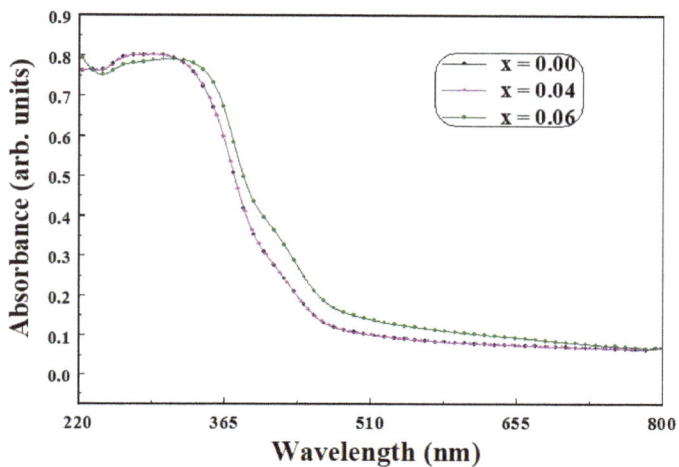

Figure 2.13 *UV-visible absorption spectra of BaTi$_{1-x}$Zr$_x$O$_3$.*

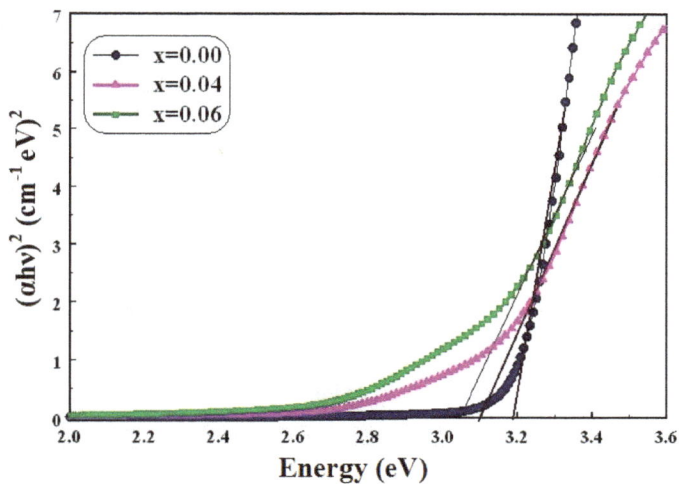

Figure 2.14 *Tauc plot of BaTi$_{1-x}$Zr$_x$O$_3$.*

Table 2.7 *Optical band gap values of BaTi$_{1-x}$Zr$_x$O$_3$.*

Samples	Band gap (eV)
x=0.00	3.181
x=0.04	3.087
x=0.06	3.043

2.3.3   Ba$_{1-x}$La$_{2x/3}$TiO$_3$

The optical band gap value has been evaluated using UV-vis data obtained in the wavelength range of 200 nm to 2000 nm. Figure 2.15 shows the UV-visible absorption spectra of Ba$_{1-x}$La$_{2x/3}$TiO$_3$. The Tauc plot of Ba$_{1-x}$La$_{2x/3}$TiO$_3$ has been drawn using the Tauc's relation [Wood and Tauc, 1972] and is given in figure 2.16. The optical band gap values for Ba$_{1-x}$La$_{2x/3}$TiO$_3$ five various La concentrations, x=0.000, 0.005, 0.015, 0.020 & 0.025 are presented in table 2.8.

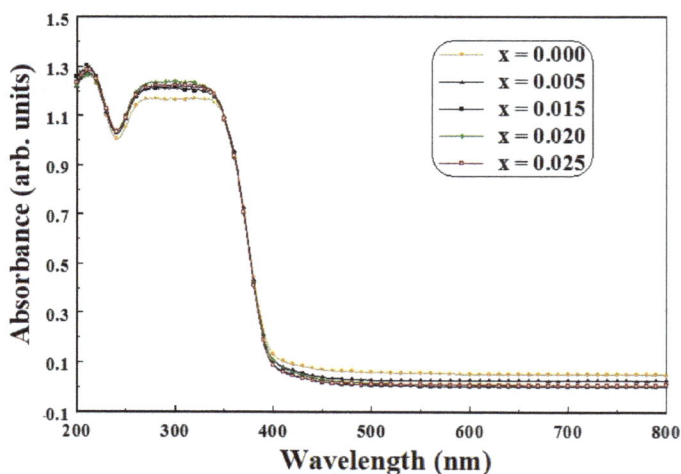

Figure 2.15 *UV-visible absorption spectra of Ba$_{1-x}$La$_{2x/3}$TiO$_3$.*

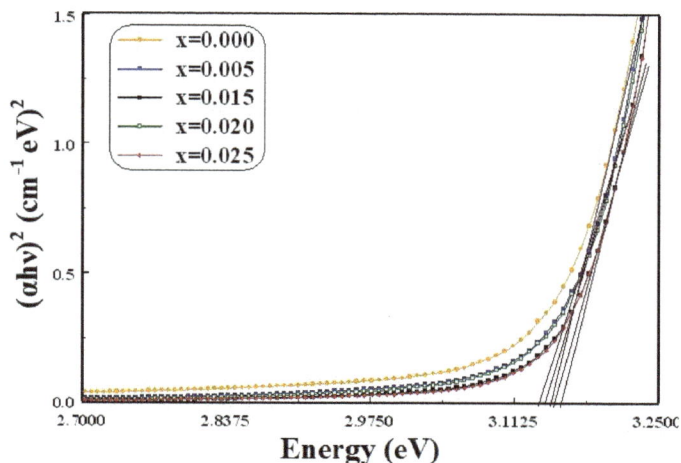

Figure 2.16 *Tauc plot of $Ba_{1-x}La_{2x/3}TiO_3$.*

Table 2.8 *Optical band gap values of $Ba_{1-x}La_{2x/3}TiO_3$.*

Samples	Band gap (eV)
x=0.000	3.146
x=0.005	3.135
x=0.015	3.156
x=0.020	3.146
x=0.025	3.147

2.3.4 BaTi$_{1-x}$Ce$_x$O$_3$

The optical band gap value has been evaluated using UV-vis data obtained in the wavelength range of 200 nm to 2000 nm. Figure 2.17 shows the UV-visible absorption spectra of BaTi$_{1-x}$Ce$_x$O$_3$. The Tauc plot of BaTi$_{1-x}$Ce$_x$O$_3$ has been drawn using the Tauc's relation [Wood and Tauc, 1972] and is given in figure 2.18. The optical band gap values for BaTi$_{1-x}$Ce$_x$O$_3$ with four different Ce concentrations, x=0.02, 0.04, 0.06 & 0.08 are presented in table 2.9.

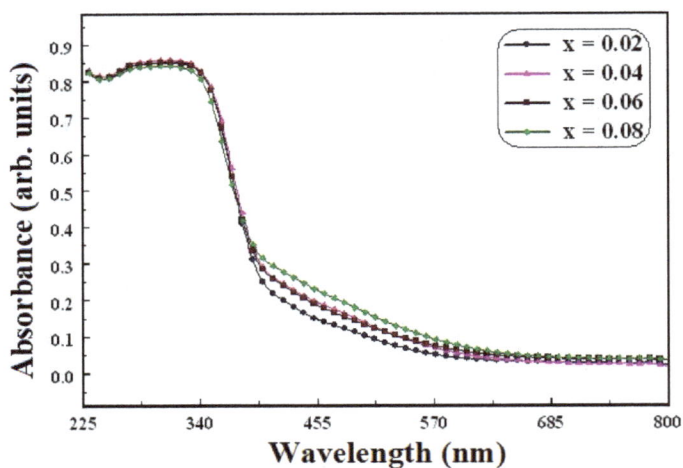

Figure 2.17 *UV-visible absorption spectra of* $BaTi_{1-x}Ce_xO_3$.

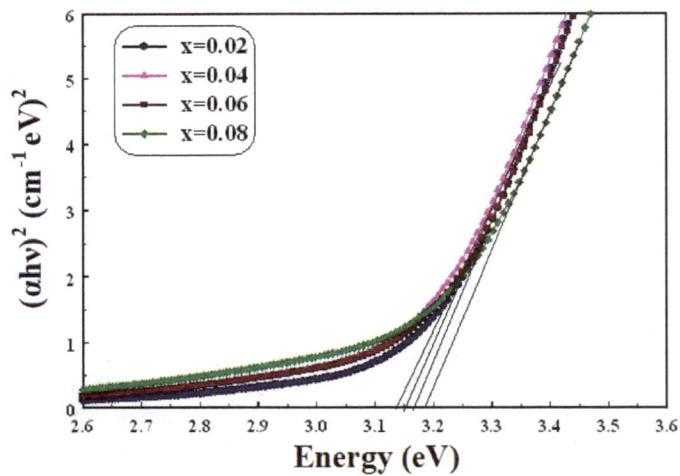

Figure 2.18 *Tauc plot of* $BaTi_{1-x}Ce_xO_3$.

Table 2.9 *Optical band gap values of* $BaTi_{1-x}Ce_xO_3$.

Samples	Band gap (eV)
x=0.02	3.165
x=0.04	3.154
x=0.06	3.136
x=0.08	3.185

2.3.5 $Ba_{1-x}Sr_xTi_{0.9}Zr_{0.1}O_3$

The optical band gap value has been evaluated using UV-vis data obtained in the wavelength range of 200 nm to 2000 nm. Figure 2.19 shows the UV-visible absorption spectra of $Ba_{1-x}Sr_xTi_{0.9}Zr_{0.1}O_3$. The Tauc plot of $Ba_{1-x}Sr_xTi_{0.9}Zr_{0.1}O_3$ has been drawn using the Tauc's relation [Wood and Tauc, 1972] and is given in figure 2.20. The optical band gap values of $Ba_{1-x}Sr_xTi_{0.9}Zr_{0.1}O_3$ for various Sr concentrations x=0.00, 0.05, 0.07 & 0.14 are presented in table 2.10.

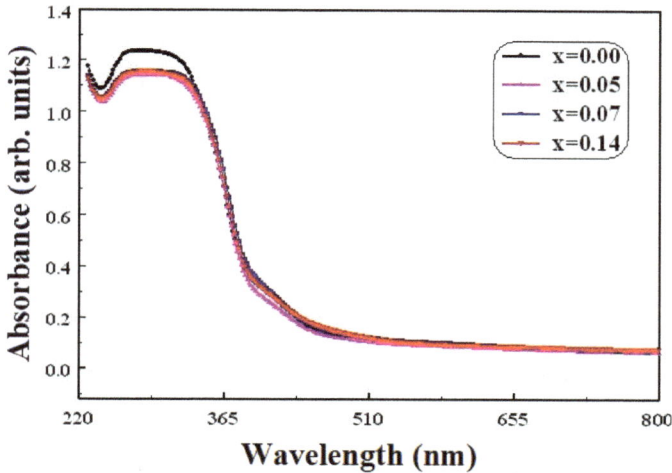

Figure 2.19 *UV-visible absorption spectra of* $Ba_{1-x}Sr_xTi_{0.9}Zr_{0.1}O_3$.

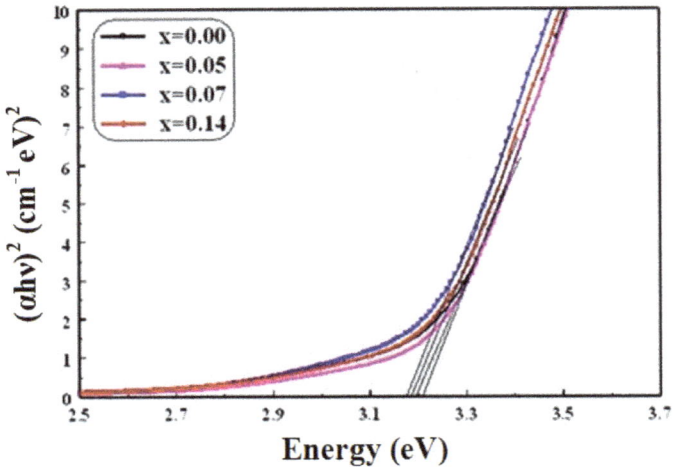

Figure 2.20 *Tauc plot of* $Ba_{1-x}Sr_xTi_{0.9}Zr_{0.1}O_3$.

Table 2.10 *Optical band gap values of* $Ba_{1-x}Sr_xTi_{0.9}Zr_{0.1}O_3$.

Samples	Band gap (eV)
x=0.00	3.208
x=0.05	3.193
x=0.07	3.174
x=0.14	3.183

2.4 Morphological characterization - SEM

2.4.1 $Ba_{1-x}Sr_xTiO_3$

The surface morphology and microstructure of $Ba_{1-x}Sr_xTiO_3$ (x=0.2, 0.4 & 0.6) have been analyzed from SEM images. The SEM images were obtained with ×1500, ×5000, ×10000 magnifications. The SEM micrographs of $Ba_{1-x}Sr_xTiO_3$ corresponding to ×10000 magnification are given in figure 2.21(a) for x=0.2, (b) for x= 0.4 & (c) for x=0.6.

(a)

(b)

(c)

Figure 2.21 *SEM images of $Ba_{1-x}Sr_xTiO_3$, (a) x=0.2, (b) x= 0.4 & (c) x=0.6.*

2.4.2 $BaTi_{1-x}Zr_xO_3$

The surface morphology and microstructure of $BaTi_{1-x}Zr_xO_3$ (x=0.00, 0.04 & 0.06) have been analyzed from SEM images. The SEM images were obtained with ×5000 to ×50000 magnifications. The SEM images of $BaTi_{1-x}Zr_xO_3$ corresponding to ×25000 magnification are given in figure 2.22 (a) for x=0.00, (b) for x= 0.04 & (c) for x=0.06.

(a)

(b)

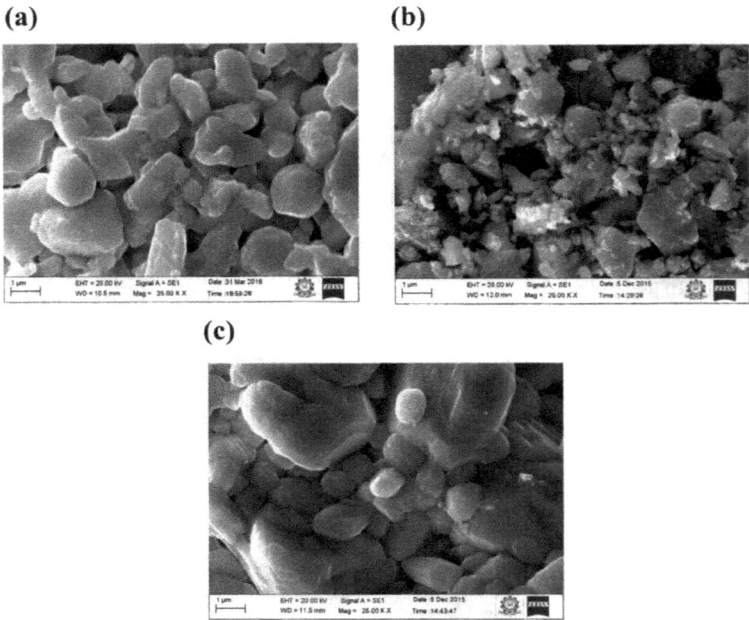

(c)

Figure 2.22 *SEM images of BaTi$_{1-x}$Zr$_x$O$_3$,* **(a)** *x=0.00,* **(b)** *x= 0.04 &* **(c)** *x=0.06.*

2.4.3 Ba$_{1-x}$La$_{2x/3}$TiO$_3$

The surface morphology and microstructure of Ba$_{1-x}$La$_{2x/3}$TiO$_3$ (x=0.000, 0.005, 0.015, 0.020 & 0.025) have been analyzed from SEM images. The SEM images were obtained with ×1500, ×5000, ×10000 magnification levels. The SEM images of Ba$_{1-x}$La$_{2x/3}$TiO$_3$ corresponding to ×10000 magnification are given in figure 2.23(a) for x=0.000, (b) for x=0.005, (c) for x=0.015, (d) for x=0.020 & (e) for x=0.025.

Figure 2.23 *SEM images of* $Ba_{1-x}La_{2x/3}TiO_3$, *(a)* $x=0.000$, *(b)* $x=0.005$, *(c)* $x=0.015$, *(d)* $x=0.020$ & *(e)* $x=0.025$.

2.4.4 BaTi$_{1-x}$Ce$_x$O$_3$

The surface morphology and microstructure of BaTi$_{1-x}$Ce$_x$O$_3$ (x=0.02, 0.04, 0.06 & 0.08) have been analyzed from scanning electron microscopy (SEM). The SEM images were obtained with ×5000 to ×50000 magnifications. The SEM images of BaTi$_{1-x}$Ce$_x$O$_3$ samples corresponding to ×10000 magnification are given in figure 2.24 (a) for x=0.02, (b) for x=0.04, (c) for x=0.06 & (d) for x=0.08.

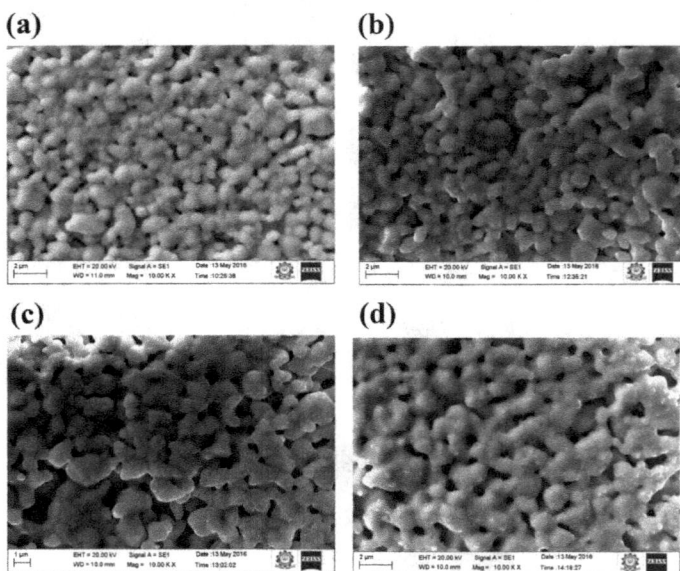

Figure 2.24 *SEM images of BaTi$_{1-x}$Ce$_x$O, (**a**) x=0.02, (**b**) x=0.04, (**c**) x=0.06 & (**d**) x=0.08.*

2.4.5 $Ba_{1-x}Sr_xTi_{0.9}Zr_{0.1}O_3$

The surface morphology and microstructure of $Ba_{1-x}Sr_xTi_{0.9}Zr_{0.1}O_3$ (x=0.00, 0.05, 0.07 & 0.14) have been analyzed from SEM images. The SEM images were obtained with ×3000 to ×20000 magnifications. The SEM images of $Ba_{1-x}Sr_xTi_{0.9}Zr_{0.1}O_3$ corresponding to ×15000 magnification are given in figure 2.25 (a) for x=0.00, (b) for x=0.05, (c) for x=0.07 & (d) for x=0.14.

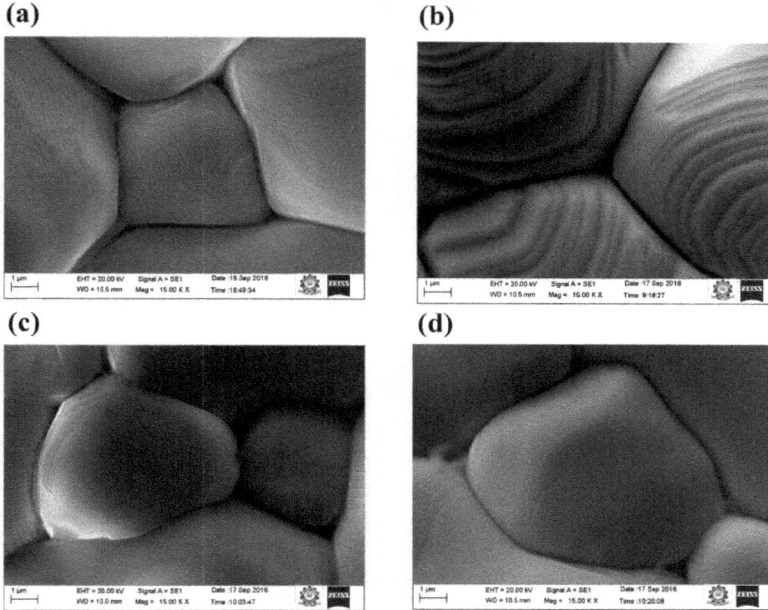

(a) **(b)** **(c)** **(d)**

Figure 2.25 *SEM images of $Ba_{1-x}Sr_xTi_{0.9}Zr_{0.1}O_3$, **(a)** x=0.00, **(b)** x=0.05, **(c)** x=0.07 & **(d)** x=0.14.*

2.5 Elemental confirmation - EDS

2.5.1 $Ba_{1-x}Sr_xTiO_3$

The chemical compositions of Sr doped $Ba_{1-x}Sr_xTiO_3$ (x=0.2, 0.4 & 0.6) ceramics have been analyzed qualitatively and quantitatively using energy dispersive X-ray spectroscopy (EDS). The EDS spectra of $Ba_{1-x}Sr_xTiO_3$ ceramic samples are shown in figure 2.26 (a) for x=0.2, (b) for x=0.4 & (c) for x=0.6. The atomic percentages and mass percentages of $Ba_{1-x}Sr_xTiO_3$ are given in table 2.11.

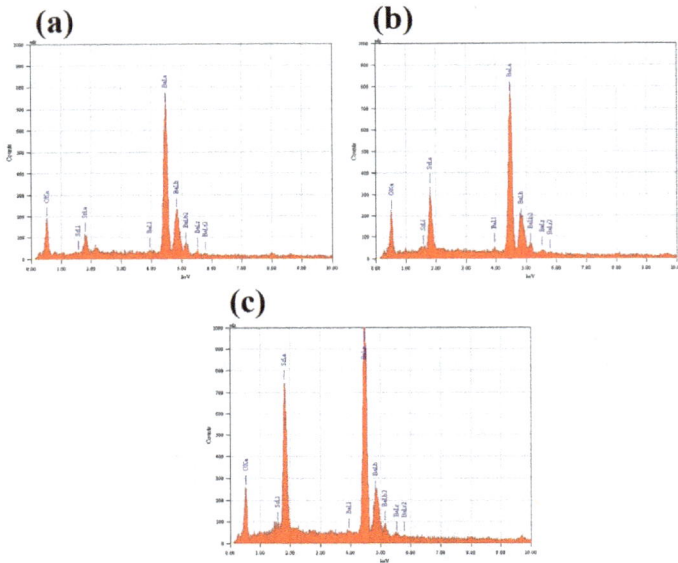

Figure 2.26 EDS spectra of $Ba_{1-x}Sr_xTiO_3$, **(a)** x=0.2, **(b)** x= 0.4 & **(c)** x=0.6.

Table 2.11 Elemental compositions of $Ba_{1-x}Sr_xTiO_3$

Sr concentration	Atomic %			Mass %		
	Ba	Sr	O	Ba	Sr	O
x=0.2	73.47	5.86	20.67	92.28	4.7	3.02
x=0.4	63.00	16.53	20.47	82.97	13.89	3.14
x=0.6	54.51	29.32	16.16	72.59	24.91	2.51

2.5.2 BaTi$_{1-x}$Zr$_x$O$_3$

The chemical compositions of Zr doped BaTi$_{1-x}$Zr$_x$O$_3$ (x=0.0, 0.04 & 0.06) ceramics have been analyzed using energy dispersive X-ray spectroscopy (EDS). The EDS spectra of the BaTi$_{1-x}$Zr$_x$O$_3$ ceramic samples are shown in figure 2.27(a) for x=0.00, (b) for x=0.04 & (c) for x=0.06. The atomic percentages and mass percentages of BaTi$_{1-x}$Zr$_x$O$_3$ are given in table 2.12.

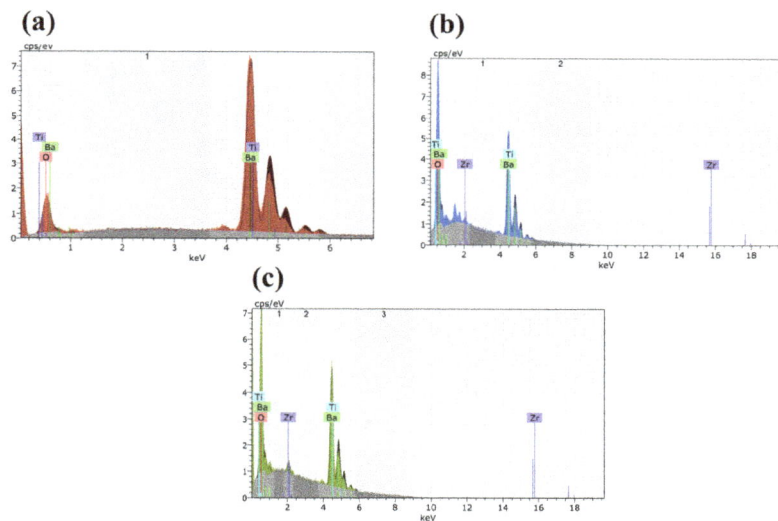

Figure 2.27 *EDS spectra of BaTi$_{1-x}$Zr$_x$O$_3$, (a) x=0.00, (b) x=0.04 & (c) x=0.06.*

Table 2.12 *Elemental compositions of BaTi$_{1-x}$Zr$_x$O$_3$.*

Zr concentration	Atomic %				Mass %			
	Ba	Ti	Zr	O	Ba	Ti	Zr	O
x=0.00	35.00	12.63	0.00	52.34	76.94	9.67	0.00	13.39
x=0.04	34.07	9.63	0.99	55.31	76.51	7.54	1.47	14.47
x=0.06	33.73	9.56	1.28	55.42	76.02	7.51	1.92	14.55

2.5.3 $Ba_{1-x}La_{2x/3}TiO_3$

The chemical compositions of La doped $Ba_{1-x}La_{2x/3}TiO_3$ (x=0.000, 0.005, 0.015, 0.020 & 0.025) ceramics have been analyzed qualitatively and quantitatively using energy dispersive X-ray spectroscopy (EDS). The EDS spectra of $Ba_{1-x}La_{2x/3}TiO_3$ ceramic samples are shown in figure 2.28 (a) for x=0.000, (b) for x=0.005, (c) for x=0.015, (d) for x=0.020 & (e) for x=0.025. The atomic percentages and mass percentages of $Ba_{1-x}La_{2x/3}TiO_3$ are given in table 2.13.

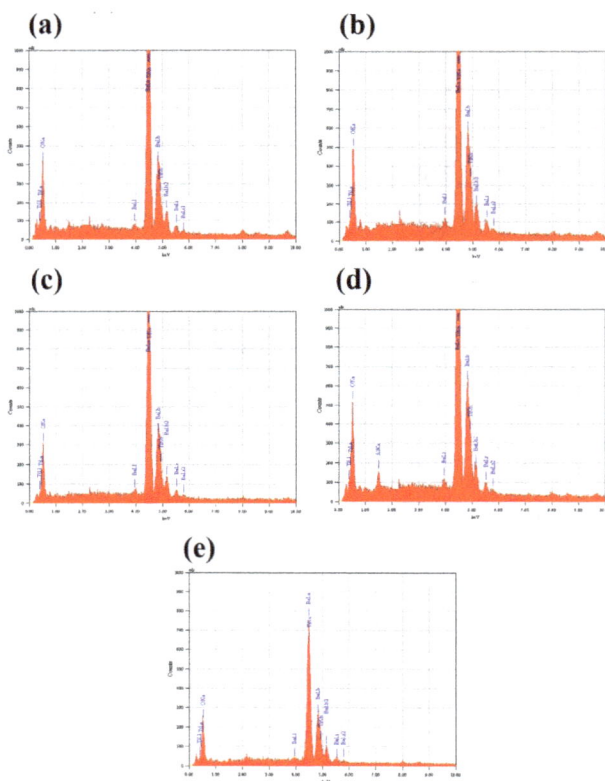

Figure 2.28 EDS spectra of $Ba_{1-x}La_{2x/3}TiO_3$, **(a)** x=0.000, **(b)** x=0.005, **(c)** x=0.015, **(d)** x=0.020 & **(e)** x=0.025.

Table 2.13 *Elemental compositions of* $Ba_{1-x}La_{2x/3}TiO_3$.

La Concentration	Atomic %			Mass %		
	Ba	**Ti**	**O**	**Ba**	**Ti**	**O**
x=0.000	86.40	9.86	3.74	58.86	19.26	21.89
x=0.005	89.90	6.82	3.28	65.35	14.22	20.44
x=0.015	92.04	4.71	3.26	68.95	10.11	20.94
x=0.020	86.55	9.52	3.37	59.44	18.75	19.84
x=0.025	91.89	3.98	4.41	66.20	8.22	25.58

2.5.4 BaTi$_{1-x}$Ce$_x$O$_3$

The chemical compositions of Ce doped $BaTi_{1-x}Ce_xO_3$ (x=0.02, 0.04, 0.06 & 0.08) ceramics have been analyzed using energy dispersive X-ray spectroscopy (EDS). The EDS spectra of all the $BaTi_{1-x}Ce_xO_3$ ceramic samples are shown in figure 2.29(a) for x=0.02, (b) for x=0.04, (c) for x=0.06 & (d) for x=0.08. The atomic percentages and mass percentages of $BaTi_{1-x}Ce_xO_3$ are given in table 2.14.

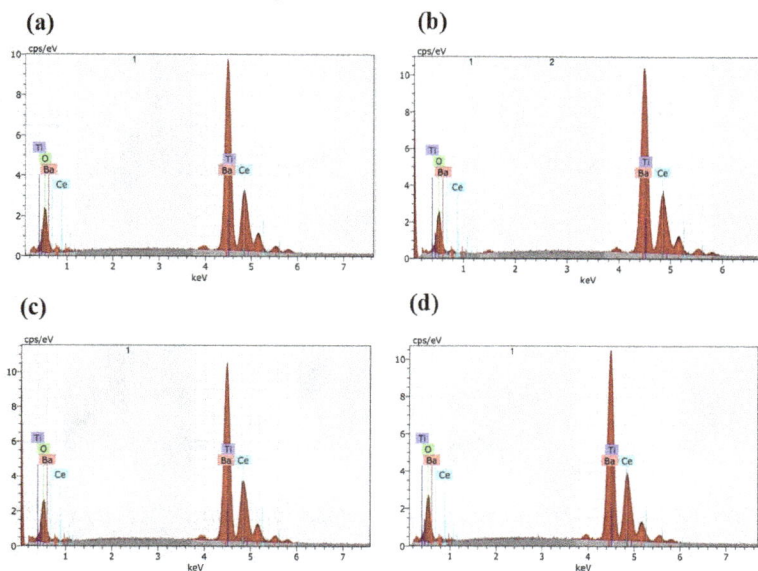

Figure 2.29 *EDS spectra of $BaTi_{1-x}Ce_xO_3$, (a) x=0.02, (b) x=0.04, (c) x=0.06 & (d) x=0.08.*

Table 2.14 *Elemental compositions of BaTi$_{1-x}$Ce$_x$O$_3$.*

Ce	Atomic %				Mass %			
concentration	Ba	Ce	Ti	O	Ba	Ce	Ti	O
x=0.02	19.56	0.38	23.74	56.59	56.34	1.10	23.57	18.99
x=0.04	18.84	0.86	23.06	57.24	54.72	2.54	23.36	19.38
x=0.06	20.32	1.01	22.03	56.61	57.02	2.88	21.59	18.51
x=0.08	19.95	2.04	22.78	55.24	54.79	5.71	21.82	17.68

2.5.5 Ba$_{1-x}$Sr$_x$Ti$_{0.9}$Zr$_{0.1}$O$_3$

The chemical compositions of Sr doped Ba$_{1-x}$Sr$_x$Ti$_{0.9}$Zr$_{0.1}$O$_3$ (x=0.00, 0.05, 0.07 & 0.14) ceramics have been analyzed using energy dispersive X-ray spectroscopy (EDS). The EDS spectra of Ba$_{1-x}$Sr$_x$Ti$_{0.9}$Zr$_{0.1}$O$_3$ ceramic samples are shown in figure 2.30 (a) for x=0.00, (b) for x=0.05, (c) for x=0.07 & (d) for x=0.14.

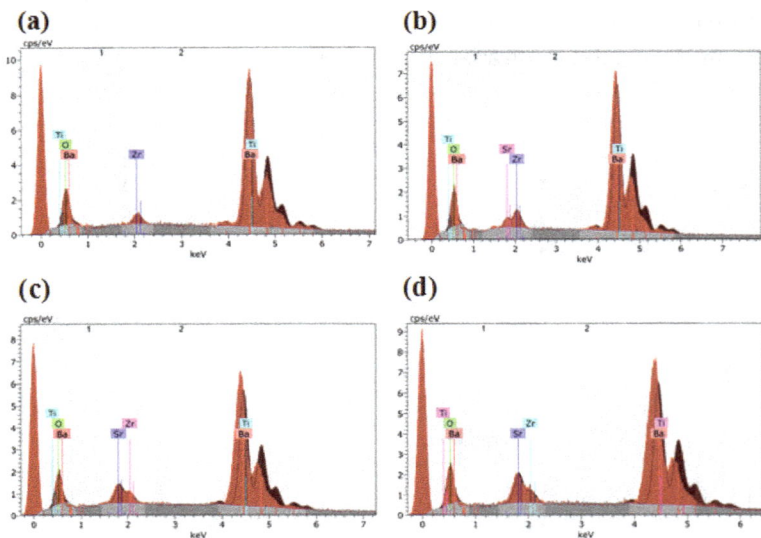

Figure 2.30 *EDS spectra of Ba$_{1-x}$Sr$_x$Ti$_{0.9}$Zr$_{0.1}$O$_3$, (**a**) x=0.00, (**b**) x=0.05, (**c**) x=0.07 & (**d**) x=0.14.*

2.6 Charge density distribution – maximum entropy method

2.6.1 $Ba_{1-x}Sr_xTiO_3$

The precise electronic structure, charge density distribution and inter-atomic bonding of the constituent atoms in the unit cell of BST have been analyzed by the maximum entropy method (MEM) [Collins, 1982] using the softwares PRIMA [Izumi, 2002] and VESTA [Momma, 2008]. Figures 2.31(a), (b) & (c) show the 3D electron density of $Ba_{1-x}Sr_xTiO_3$ with different Sr concentrations, x=0.2, 0.4 & 0.6 respectively. Figure 2.32 shows the enlarged 3D view of Ba-O and Ti-O bonds of $Ba_{1-x}Sr_xTiO_3$ for (a) x=0.2, (b) x=0.4 & (c) x=0.6. The 3D unit cell of BST with the (100) plane shaded is shown in figure 2.33(a) and the 2D contour maps of BST with enlarged view of the Ba-O bond for various Sr concentrations, x=0.2, 0.4 & 0.6 are given in figures 2.33 (b), (c) & (d) respectively. The 3D unit cell of BST with the (200) plane shaded is shown in figure 2.34(a) and the 2D contour maps of BST with enlarged view of the Ti-O bond for various Sr concentrations x=0.2, 0.4 & 0.6 are given in figures 2.34 (b), (c) & (d) respectively. The 3D unit cell of BST with the (101) plane shaded is shown in the figure 2.35(a) and the 2D contour maps of BST with enlarged view of the Ti-O bond for various Sr concentrations x=0.2, 0.4 & 0.6 are given in figures 2.35 (b), (c) & (d) respectively. The 1D line profiles of the Ba-O and the Ti-O bonds are shown in figures 2.36 and 2.37 respectively. The bond lengths and mid bond density values of Ba-O and Ti-O bonds are given in table 2.15.

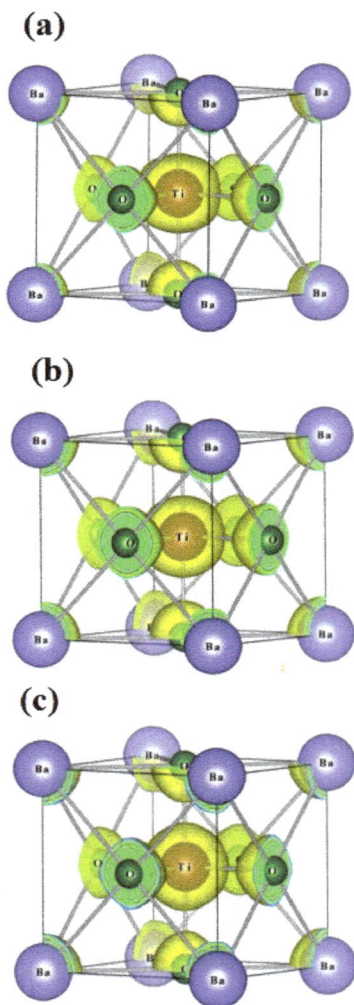

Figure 2.31 *3D electron density of* $Ba_{1-x}Sr_xTiO_3$, *(a)* x=0.2, *(b)* x=0.4 & *(c)* x=0.6 *(iso-surface level : 0.8 e/Å^3).*

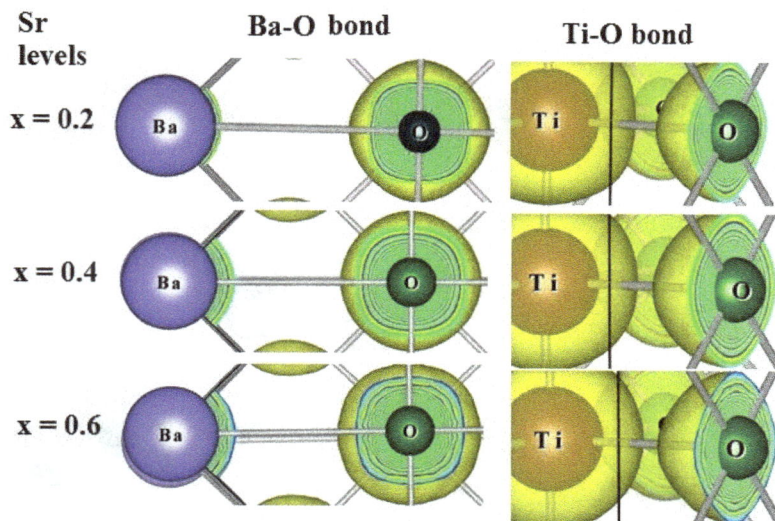

Figure 2.32 *Enlarged 3D view of Ba-O and Ti-O bonds (iso-surface level: 0.8 e/Å³).*

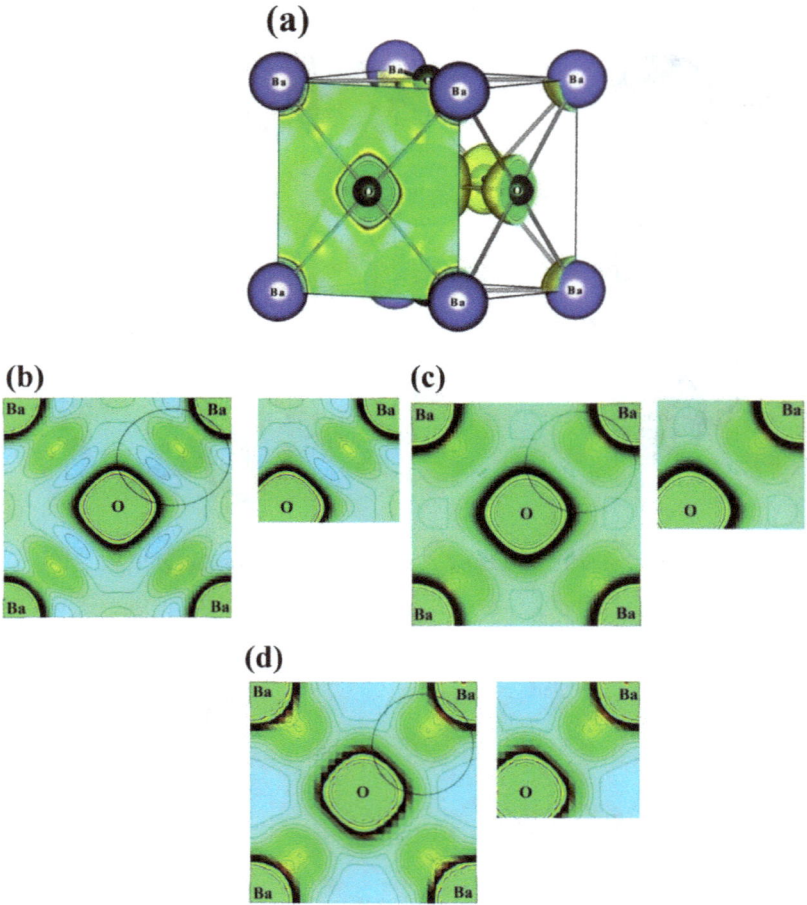

Figure 2.33(a) *3D unit cell of BST with the (100) plane shaded, 2D electron density contour maps of $Ba_{1-x}Sr_xTiO_3$ with enlarged view of the Ba-O bond, (b) x=0.2, (c) x=0.4 & (d) x=0.6 (contour range: 0 $e/Å^3$ to 0.8 $e/Å^3$, contour interval: 0.04 $e/Å^3$).*

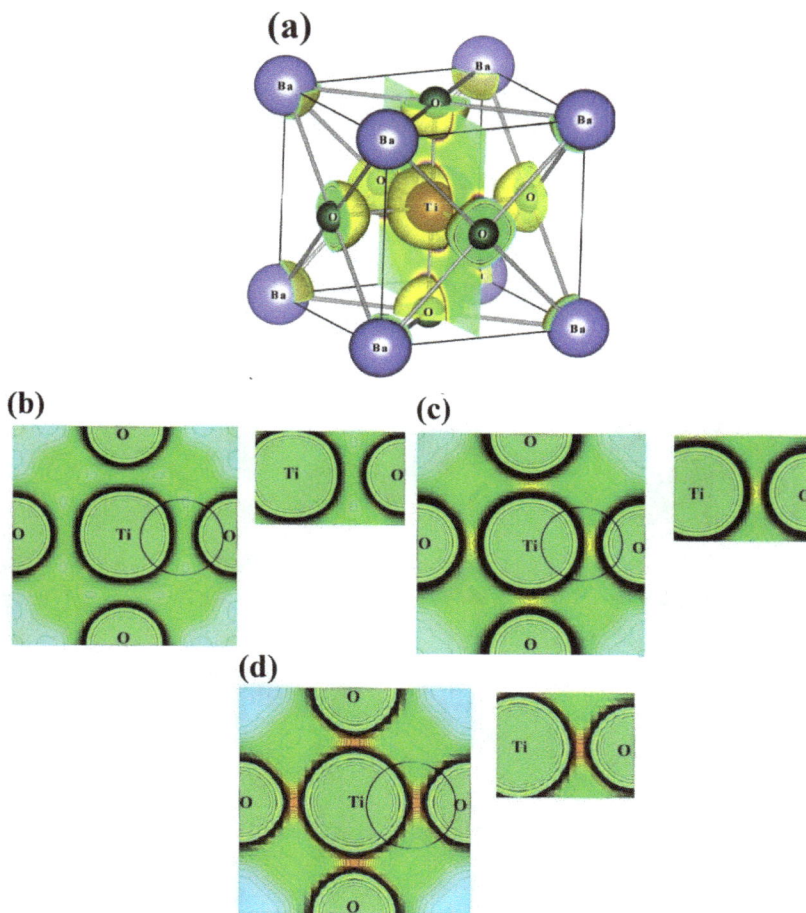

Figure 2.34(a) *3D unit cell of BST with the (200) plane shaded, 2D electron density contour maps of $Ba_{1-x}Sr_xTiO_3$ with enlarged view of the Ti-O bond,* **(b)** *x=0.2,* **(c)** *x=0.4 & (d) x=0.6 (contour range: 0 $e/Å^3$ to 0.8 $e/Å^3$, contour interval: 0.04 $e/Å^3$).*

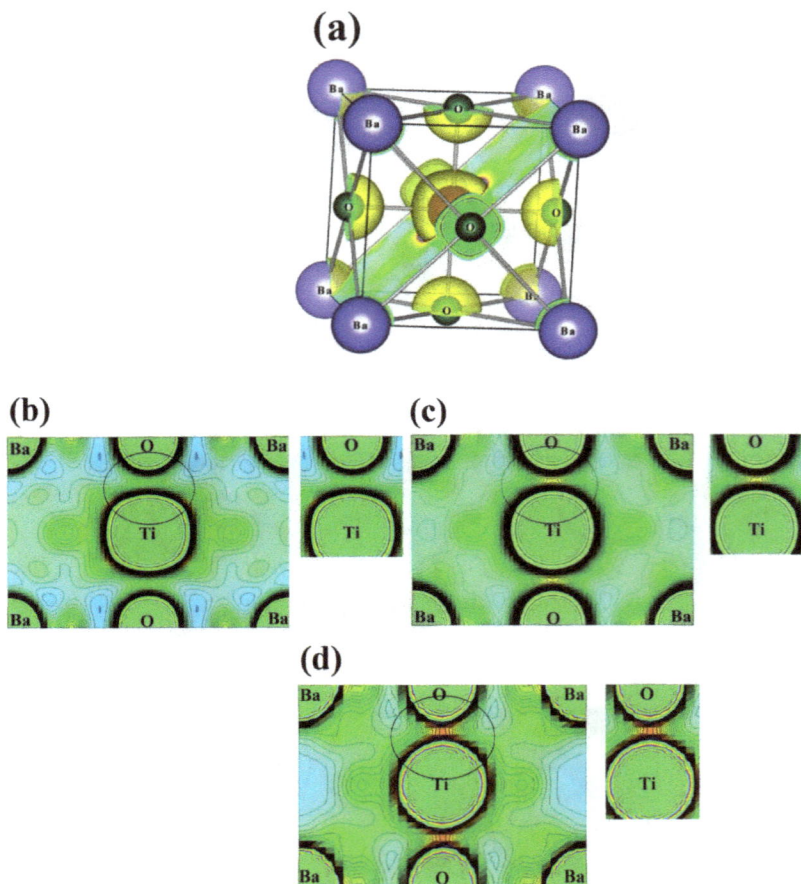

Figure 2.35(a) *3D unit cell of BST with the (101) plane shaded, 2D electron density contour maps of $Ba_{1-x}Sr_xTiO_3$ with enlarged view of the Ti-O bond,* **(b)** *x=0.2,* **(c)** *x=0.4 &* **(d)** *x=0.6 (contour range: 0 e/$Å^3$ to 0.8 e/$Å^3$, contour interval: 0.04 e/$Å^3$).*

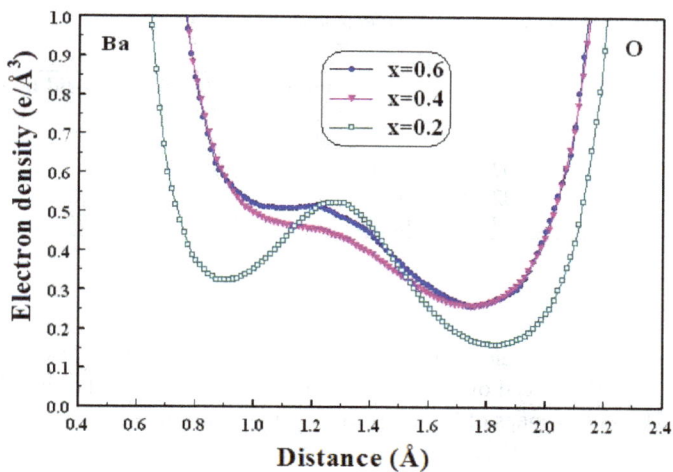

Figure 2.36 *1D line profiles of Ba-O bond.*

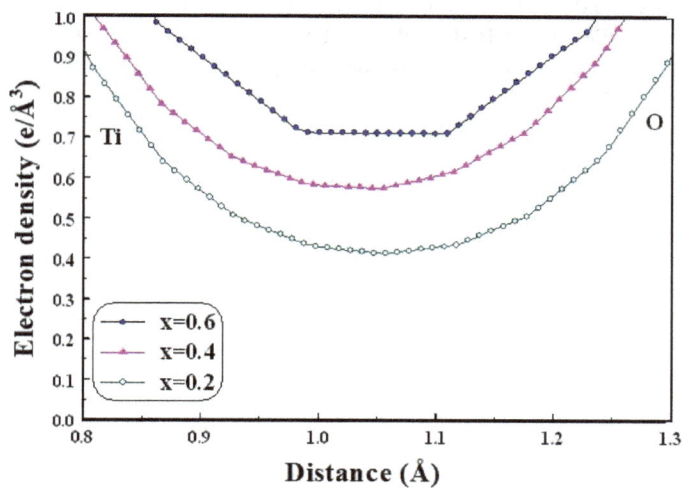

Figure 2.37 *1D line profiles of Ti-O bond.*

Table 2.15 *Bond lengths and mid bond electron density values of $Ba_{1-x}Sr_xTiO_3$.*

Sr concentration	Ba-O		Ti-O	
	Bond length (Å)	Mid bond density (e/Å³)	Bond length (Å)	Mid bond density (e/Å³)
x=0.2	2.807	0.515	1.985	0.413
x=0.4	2.802	0.453	1.981	0.574
x=0.6	2.787	0.508	1.971	0.709

2.6.2 BaTi$_{1-x}$Zr$_x$O$_3$

Figure 2.38 shows the 3D electron density of BaTi$_{1-x}$Zr$_x$O$_3$ for (a) x=0.00, (b) x= 0.04 & (c) x=0.06. The 3D unit cell of BZT with the (100) plane shaded is shown in figure 2.39 (a) and the 2D contour maps of BZT with enlarged view of the Ba-O bond for various Zr concentrations x=0.00, 0.04 & 0.06 are given in figures 2.39 (b), (c) & (d) respectively. The 3D unit cell of BZT with the (200) plane shaded is shown in figure 2.40(a) and the 2D contour maps of BZT with enlarged view of the Ti-O bond for various Zr concentrations x=0.00, 0.04 & 0.06 are given in figures 2.40 (b), (c) & (d) respectively. The 1D line profiles of the Ba-O and Ti-O bonds are shown in figures 2.41 and 2.42 respectively. The bond lengths and electron density values of Ba-O and Ti-O bonds are given in table 2.16.

(a)

(b)

(c)

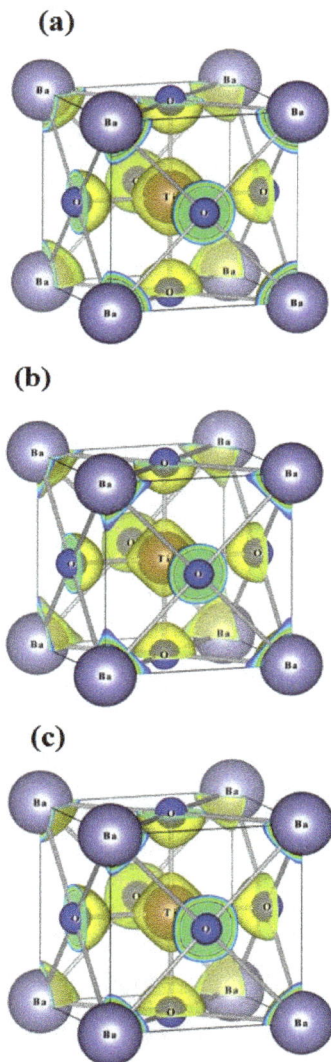

Figure 2.38 *3D electron density of $BaTi_{1-x}Zr_xO_3$, **(a)** x=0.00, **(b)** x=0.04 & **(c)** x=0.06 (iso-surface level: 1 $e/Å^3$).*

(a)

(b)

(c)

(d)

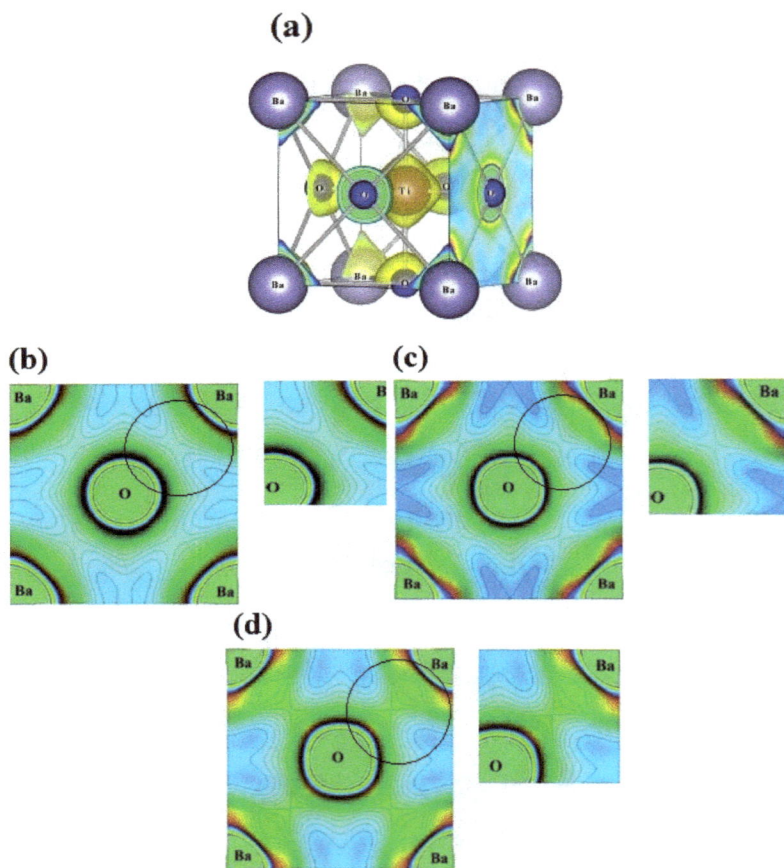

Figure 2.39(a) *3D unit cell of BZT with the (100) plane shaded, 2D electron density contour maps of BaTi$_{1-x}$Zr$_x$O$_3$ with enlarged view of the Ba-O bond, **(b)** x=0.00, **(c)** x=0.04 & **(d)** x=0.06 (contour range: 0 e/Å^3 to 1 e/Å^3, contour interval: 0.04 e/Å^3).*

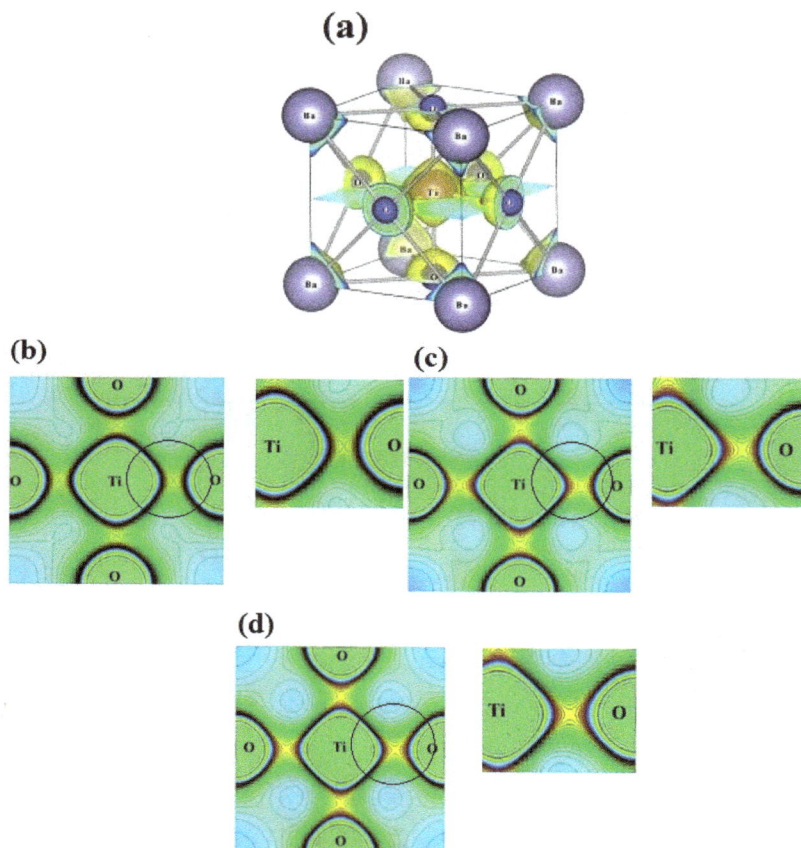

Figure 2.40 *(a) 3D unit cell of BZT with the (200) plane shaded, 2D electron density contour maps of BaTi$_{1-x}$Zr$_x$O$_3$ with enlarged view of the Ti-O bond, (b) x=0.00, (c) x=0.04 & (d) x=0.06 (contour range: 0 e/Å^3 to 1 e/Å^3, contour interval: 0.04 e/Å^3)*

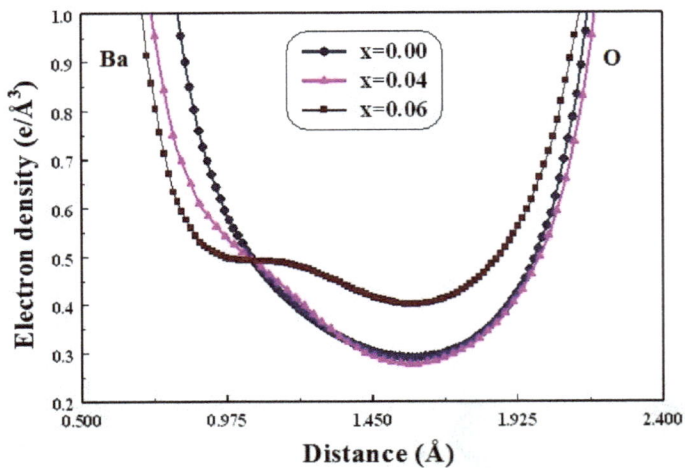

Figure 2.41 *1D line profiles of Ba-O bond.*

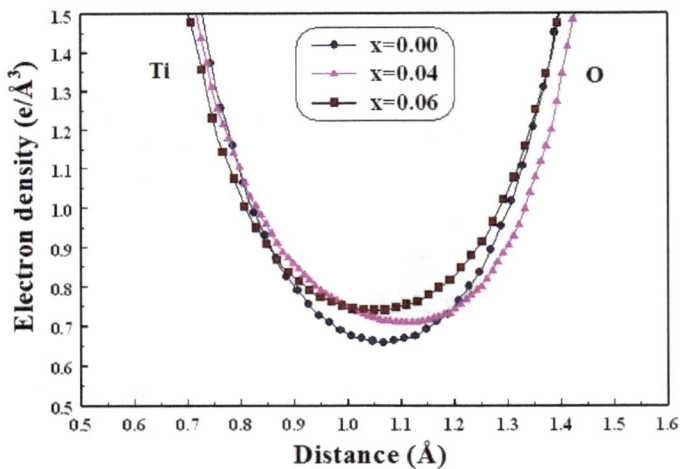

Figure 2.42 *1D line profiles of Ti-O bond.*

Table 2.16 *Bond lengths and mid bond electron density values of* $BaTi_{1-x}Zr_xO_3$.

Zr concentration	Ba-O			Ti-O		
	Bond length (Å)	Distance (Å)	Mid bond density (e/Å^3)	Bond length (Å)	Distance (Å)	Mid bond density (e/Å^3)
x=0.00	2.833	1.566	0.291	2.003	1.067	0.657
x=0.04	2.841	1.570	0.277	2.008	1.130	0.708
x=0.06	2.843	1.600	0.401	2.010	1.060	0.738

2.6.3 $Ba_{1-x}La_{2x/3}TiO_3$

Figures 2.43(a), (b), (c), (d) & (e) show the 3D electron density distributions of $Ba_{1-x}La_{2x/3}TiO_3$ with different La concentrations, x=0.000, 0.005, 0.015, 0.020 & 0.025 respectively. The 3D unit cell of BLT with the (001) plane shaded is shown in figure 2.44(a) and the 2D contour maps of BLT with enlarged view of the Ba-O1 bond for various La concentrations x=0.000, 0.005, 0.015, 0.020 & 0.025 are given in figures 2.44(b), (c), (d), (e) & (f) respectively. The 3D unit cell of BLT with the (002) plane shaded is shown in figure 2.45(a) and the 2D contour maps of BLT with enlarged view of the Ti-O2 bond for various La concentrations, x=0.000, 0.005, 0.015, 0.020 & 0.025 are given in figures 2.45 (b), (c), (d), (e) & (f) respectively. The 1D line profiles of the Ba-O1 and Ti-O2 bonds are shown in figures 2.46 and 2.47 respectively. The bond lengths and electron density values of Ba-O1 and Ti-O2 bonds are given in table 2.17.

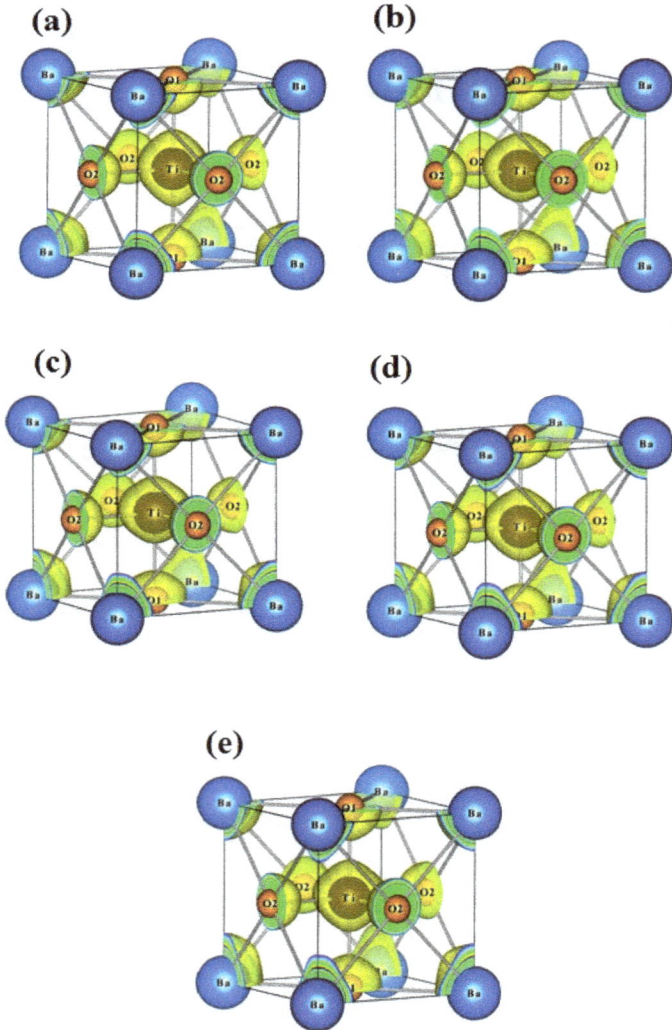

Figure 2.43 *3D electron density of* $Ba_{1-x}La_{2x/3}TiO_3$ *(a)* $x=0.000$, *(b)* $x=0.005$, *(c)* $x=0.015$, *(d)* $x=0.020$ & *(e)* $x=0.025$ *(iso-surface level: 1 e/$Å^3$).*

(a)

(b)

(c)

(d)

(e)

(f)

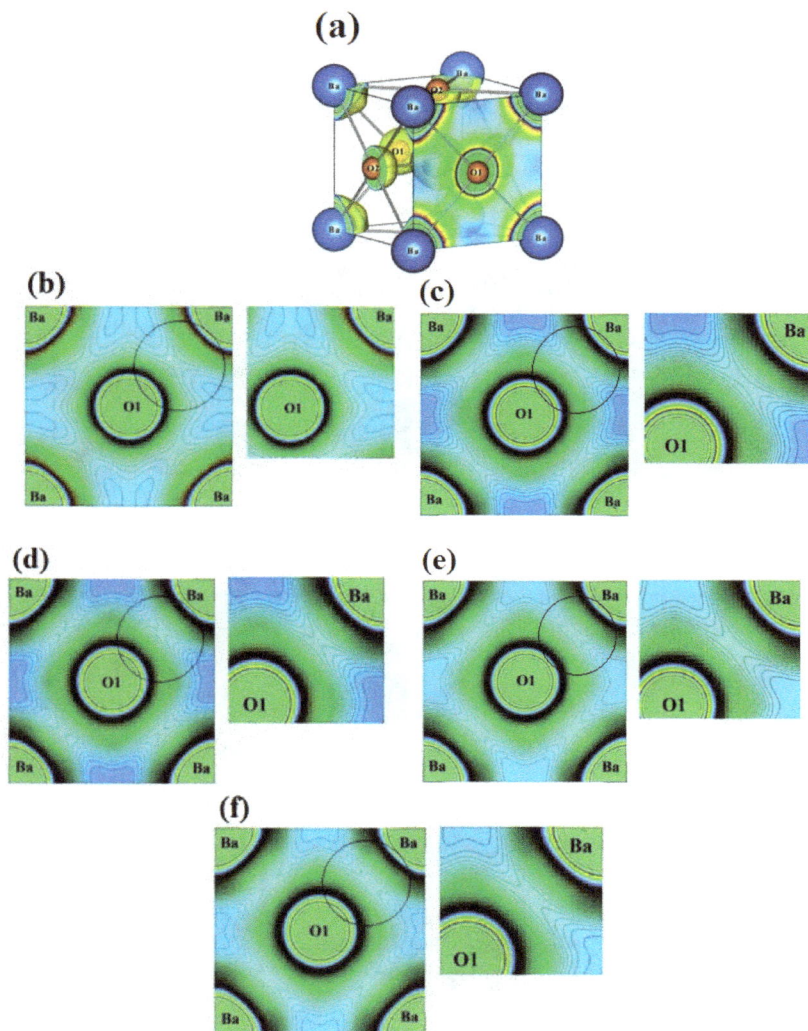

Figure 2.44(a) *3D unit cell of BLT with the (001) plane shaded, 2D electron density contour maps of $Ba_{1-x}La_{2x/3}TiO_3$ with enlarged view of the Ba-O1 bond, **(b)** x=0.000, **(c)** x=0.005, **(d)** x=0.015, **(e)** x=0.020 & **(f)** x=0.025 (contour range: 0 e/Å³ to 1 e/Å³, contour interval: 0.03 e/Å³).*

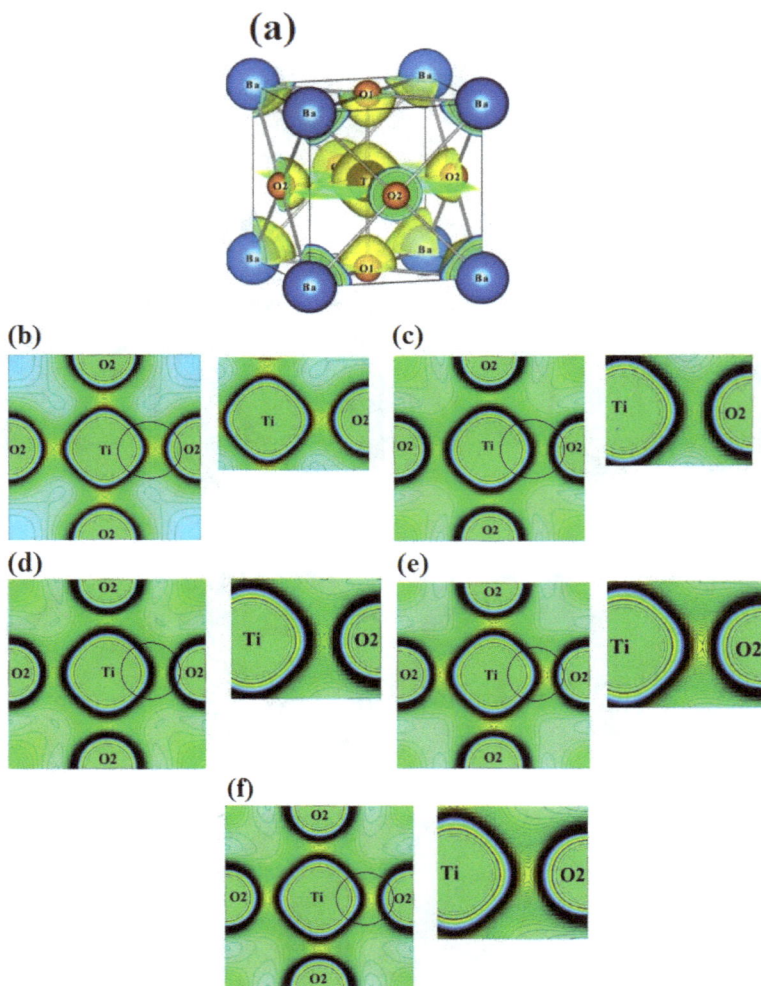

Figure 2.45(a) *3D unit cell of BLT with the (002) plane shaded, 2D electron density contour maps of $Ba_{1-x}La_{2x/3}TiO_3$ with enlarged view of the Ti-O2 bond, **(b)** x=0.000, **(c)** x=0.005, **(d)** x=0.015, **(e)** x=0.020 & **(f)** x=0.025 (contour range: 0 e/$Å^3$ to 1 e/$Å^3$, contour interval: 0.03 e/$Å^3$).*

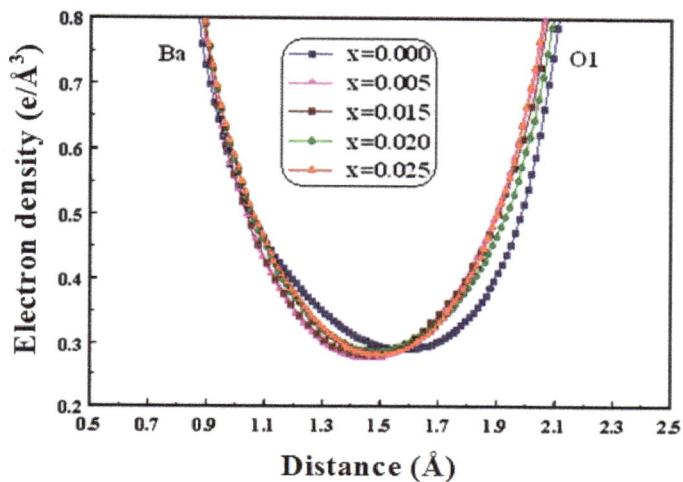

Figure 2.46 *1D line profiles of Ba-O1 bond.*

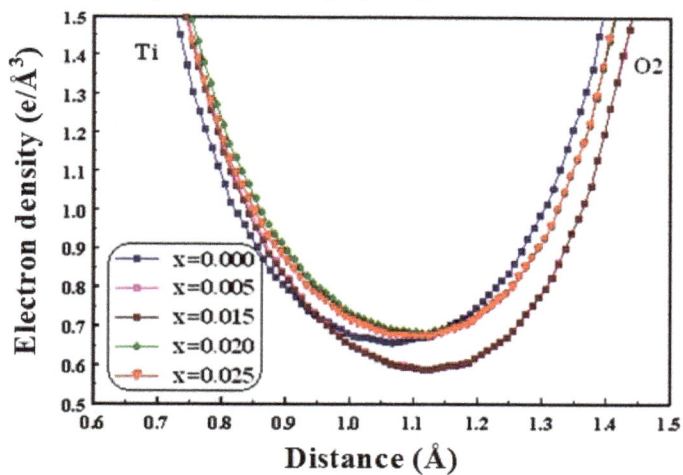

Figure 2.47 *1D line profiles of Ti-O2 bond.*

Table 2.17 *Bond lengths and mid bond electron density values of $Ba_{1-x}La_{2x/3}TiO_3$.*

La Concentration	Ba-O1		Ti-O2	
	Bond length (Å)	Mid bond density (e/Å³)	Bond length (Å)	Mid bond density (e/Å³)
x=0.000	2.833	0.309	2.003	0.674
x=0.005	2.829	0.278	2.000	0.649
x=0.015	2.828	0.283	1.999	0.659
x=0.020	2.825	0.293	1.998	0.742
x=0.025	2.824	0.288	1.997	0.724

2.6.4 $BaTi_{1-x}Ce_xO_3$

Figures 2.48(a), (b), (c) & (d) show the 3D electron density of $BaTi_{1-x}Ce_xO_3$ with various Ce concentrations x=0.02, 0.04, 0.06, 0.08 respectively. The 3D unit cell of BCT with the (100) plane shaded is shown in figure 2.49 (a) and the 2D contour maps of BCT with enlarged view of the Ba-O bond for various Ce concentrations x=0.02, 0.04, 0.06 & 0.08 are given in figures 2.49 (b), (c), (d) & (e) respectively. The 3D unit cell of BCT with the (200) plane shaded is shown in figure 2.50(a) and the 2D contour maps of BCT with enlarged view of the Ti-O bond for various Ce concentrations x=0.02, 0.04, 0.06 & 0.08 are given in figures 2.50 (b), (c), (d) & (e) respectively. The 1D line profiles of the Ba-O and Ti-O bonds are shown in figures 2.51 and 2.52 respectively. The bond lengths and electron density values of the Ba-O and Ti-O bonds are given in table 2.18.

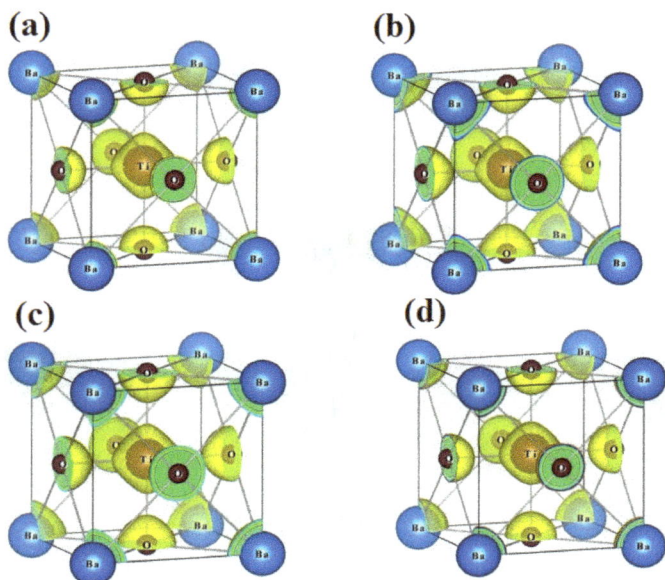

Figure 2.48 *3D electron density of BaTi$_{1-x}$Ce$_x$O$_3$* **(a)** *x=0.02,* **(b)** *x=0.04,* **(c)** *x=0.06 &* **(d)** *x=0.08 (iso-surface level: 1 e/Å^3).*

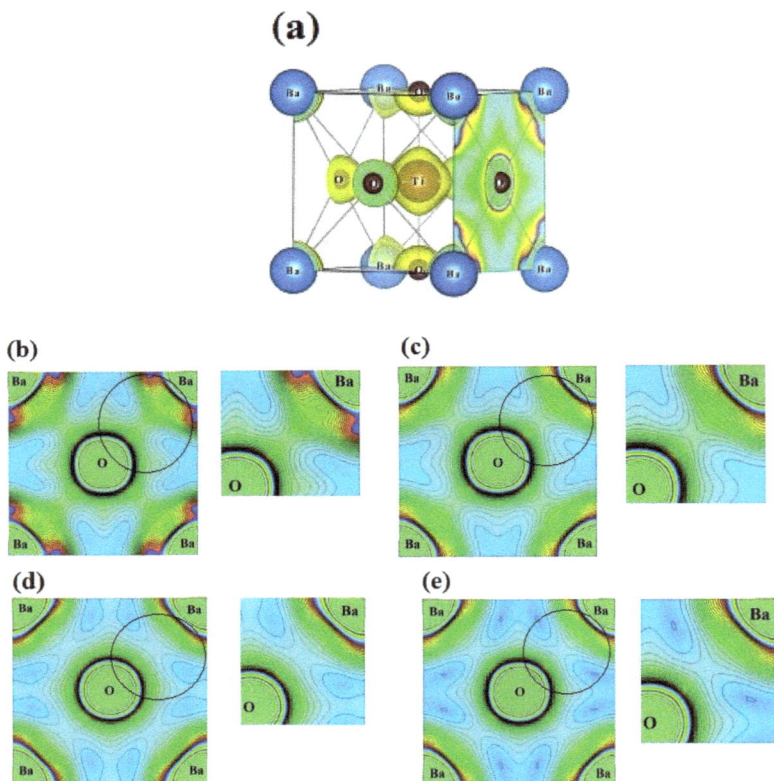

Figure 2.49(a) *3D unit cell of BCT with the (100) plane shaded, 2D electron density contour maps of* $BaTi_{1-x}Ce_xO_3$ *with enlarged view of the Ba-O bond,* **(b)** *x=0.02,* **(c)** *x=0.04,* **(d)** *x=0.06 &* **(e)** *x=0.08 (contour range: 0* $e/Å^3$ *to 1* $e/Å^3$, *contour interval: 0.04* $e/Å^3$)

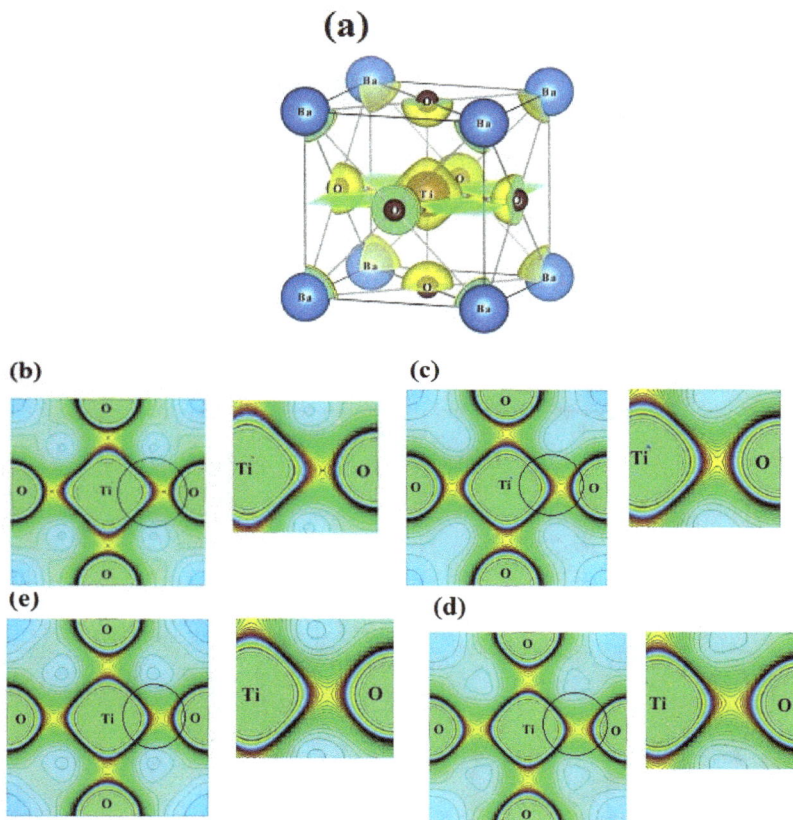

Figure 2.50 (a) 3D unit cell of BCT with the (200) plane shaded, 2D electron density contour maps of $BaTi_{1-x}Ce_xO_3$ with enlarged view of the Ti-O bond, *(b)* x=0.02, *(c)* x=0.04, *(d)* x=0.06 & *(e)* x=0.08(contour range: 0 e/$\mathring{A}^3$ to 1 e/$\mathring{A}^3$, contour interval: 0.04 e/$\mathring{A}^3$)

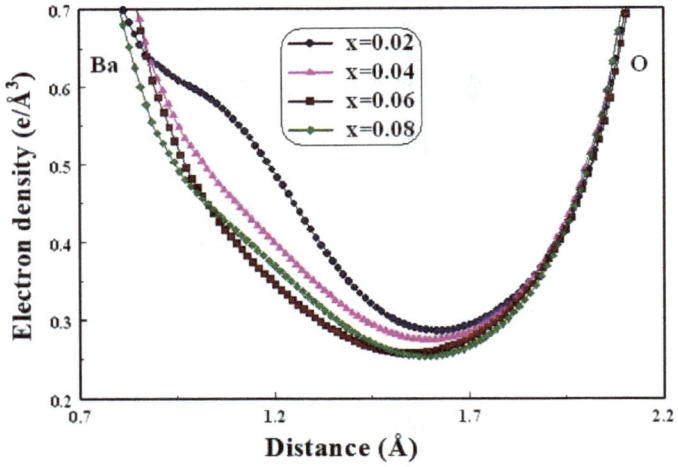

Figure 2.51 1D line profiles of Ba-O bond.

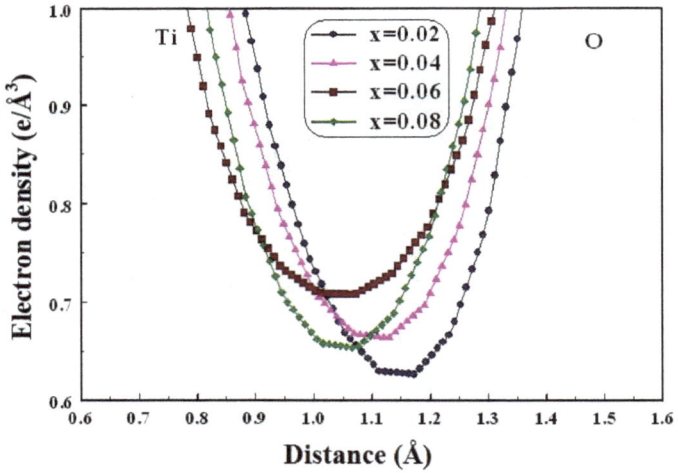

Figure 2.52 1D line profiles of Ti-O bond.

Table 2.18 *Bond lengths and mid bond electron density values of* $BaTi_{1-x}Ce_xO_3$.

Ce concentration	Ba-O		Ti-O	
	Bond length (Å)	Mid bond density $(e/\text{Å}^3)$	Bond length (Å)	Mid bond density $(e/\text{Å}^3)$
x=0.02	2.797	0.286	1.978	0.626
x=0.04	2.839	0.274	2.007	0.664
x=0.06	2.852	0.257	2.017	0.707
x=0.08	2.858	0.257	2.021	0.653

2.6.5 $Ba_{1-x}Sr_xTi_{0.9}Zr_{0.1}O_3$

Figures 2.53(a), (b), (c) & (d) show the 3D electron density of $Ba_{1-x}Sr_xTi_{0.9}Zr_{0.1}O_3$ with various Sr concentrations x=0.00, 0.05, 0.07, 0.14 respectively. The 3D unit cell of BSZT with the (100) plane shaded is shown in figure 2.54(a) and the 2D contour maps of BSZT with enlarged view of the Ba-O bond for various Sr concentrations x=0.00, 0.05, 0.07 & 0.14 are given in figures 2.54(b), (c), (d) & (e) respectively. The 3D unit cell of BSZT with the (200) plane shaded is shown in the figure 2.55(a) and the 2D contour maps of BSZT with enlarged view of the Ti-O bond for various Sr concentrations x=0.00, 0.05, 0.07 & 0.14 are given in figures 2.51 (b), (c), (d) & (e) respectively. The 1D line profiles of the Ba-O and Ti-O bonds are shown in figures 2.56 and 2.57 respectively. The bond lengths and electron density values of the Ba-O and Ti-O bonds are given in table 2.19.

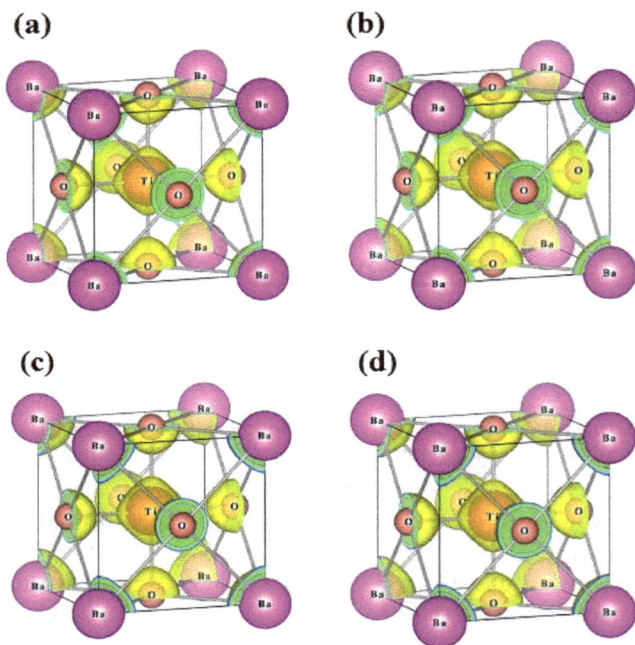

Figure 2.53 *3D electron density of $Ba_{1-x}Sr_xTi_{0.9}Zr_{0.1}O_3$ (a) x=0.00, (b) x=0.05, (c) x= 0.07 & (d) x= 0.14 (iso-surface level: 1 e/Å³).*

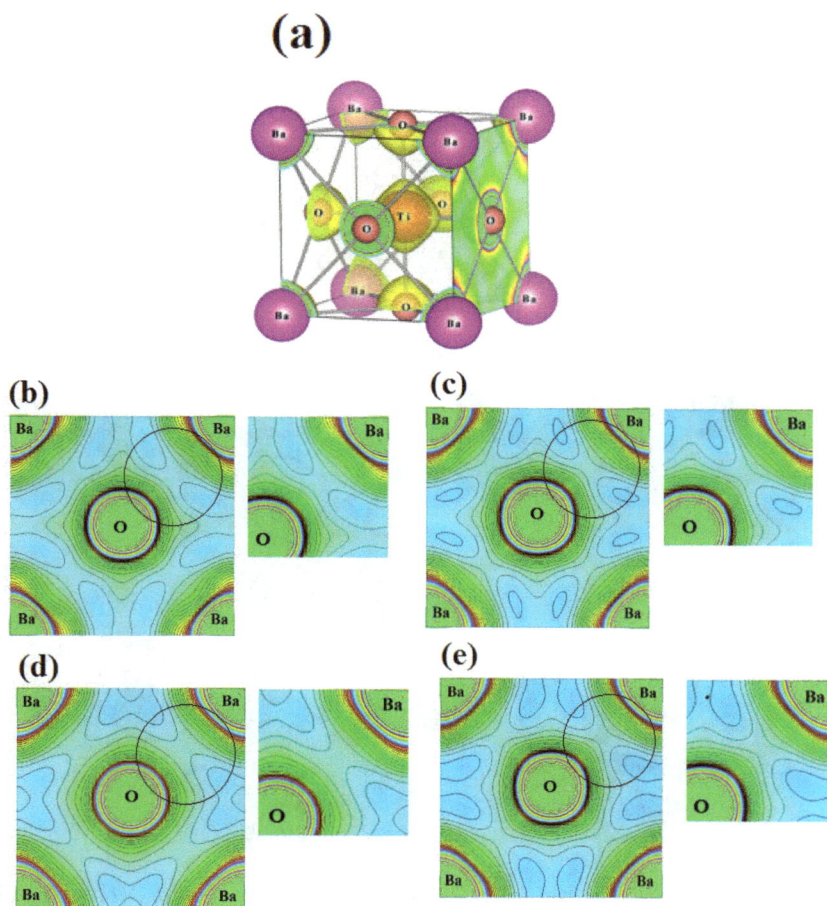

Figure 2.54(a) *3D unit cell of BSZT with the (100) plane shaded, 2D electron density contour maps of $Ba_{1-x}Sr_xTi_{0.9}Zr_{0.1}O_3$ with enlarged view of the Ba-O bond, **(b)** x=0.00, **(c)** x=0.05, **(d)** x= 0.07 & **(e)** x= 0.14 (contour range: 0 e/$Å^3$ to 1 e/$Å^3$, contour interval: 0.04 e/$Å^3$)*

.

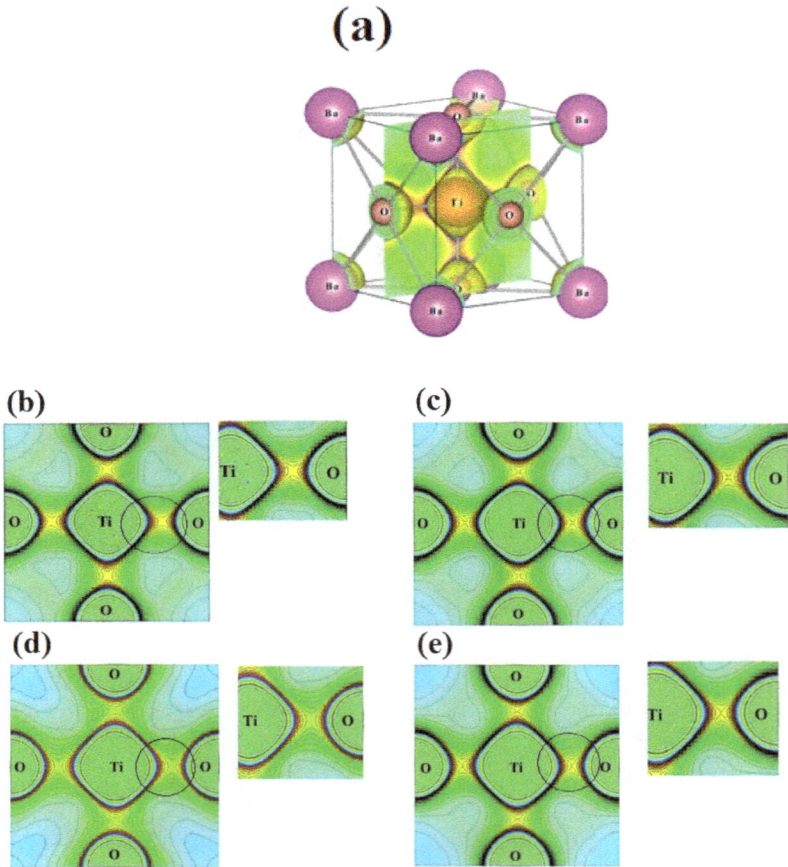

Figure 2.55 (a) *3D unit cell of BSZT with the (200) plane shaded, 2D electron density contour maps of* $Ba_{1-x}Sr_xTi_{0.9}Zr_{0.1}O_3$ *with enlarged view of the Ti-O bond,* **(b)** *x=0.00,* **(c)** *x=0.05,* **(d)** *x= 0.07 &* **(e)** *x= 0.14 (contour range: 0 e/$Å^3$ to 1 e/$Å^3$ contour interval: 0.04 e/$Å^3$).*

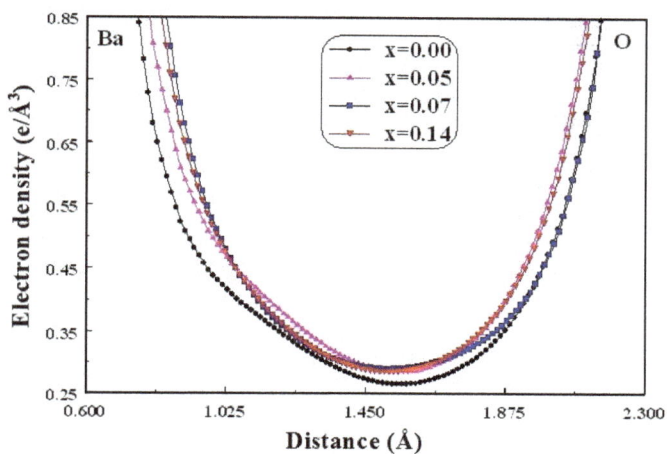

Figure 2.56 1D line profiles of Ba-O bond.

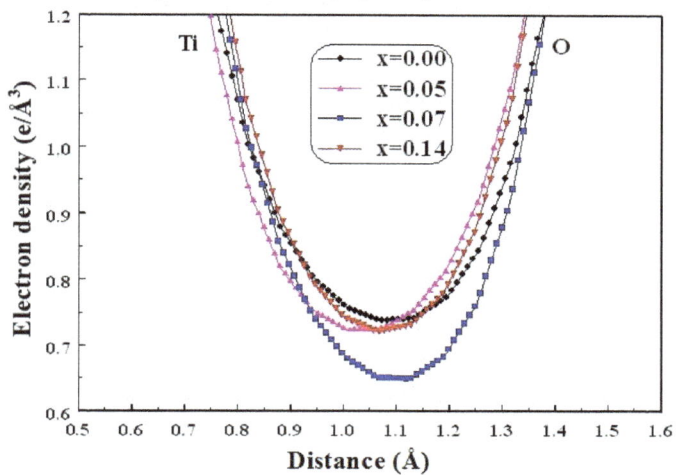

Figure 2.57 1D line profiles of Ti-O bond.

117

Table 2.19 *Bond lengths and mid bond electron density values of $Ba_{1-x}Sr_xTi_{0.9}Zr_{0.1}O_3$.*

Sr concentration	Ba-O		Ti-O	
	Bond length (Å)	Mid bond density (e/Å³)	Bond length (Å)	Mid bond density (e/Å³)
x=0.00	2.850	0.278	2.015	0.761
x=0.05	2.846	0.302	2.012	0.726
x=0.07	2.837	0.296	2.006	0.680
x=0.14	2.832	0.294	2.002	0.741

References

[1]   Collins D. M, Nature, 298, 49 (1982). https://doi.org/10.1038/298049a0

[2]   Izumi F, Dilanien R.A, Recent Research Developments in Physics Part II, Vol.3, Transworld Research Network. Trivandrum, 699–726, (2002).

[3]   Momma K, Izumi F, VESTA: a three-dimensional visualization system for electronic and structural analysis. J. Appl. Crystallogr. 41, 653 (2008). https://doi.org/10.1107/S0021889808012016

[4]   Rietveld H. M, J. Appl. Crystallogr. 2, 65 (1969). https://doi.org/10.1107/S0021889869006558

[5]   Wood D. L, Tauc J, Phys Rev B. 5, 3144 (1972). https://doi.org/10.1103/PhysRevB.5.3144

[6]   Petricek V, Dusek M, Palatinus L, Kristallogr Z, Crystallographic Computing System JANA2006: General features, 229, 345 (2014).

Chapter 3

Analysis of Results

Abstract

Chapter 3 provides the detailed analysis of results obtained for all the doped titanates from various characterization methods and analytical procedures described in chapter 1. The preparation details of samples of ceramic systems chosen for this research work are presented in section 3.2. Section 3.3 gives the structural analysis of the prepared ceramic using powder X-ray diffraction, section 3.4 gives the grain size analysis and section 3.5 explains the optical characterization by UV-visible analysis. The morphological and microstructural analysis by SEM images is given in section 3.6, the chemical compositions of all the prepared ceramic systems have been analyzed in section 3.7. The precise electronic structure, inter-atomic bonding and electron density distribution in the unit cell of the prepared ceramic solid solutions are explained in section 3.8.

Keywords

Analysis, X-Ray, Microstructure, Optical Band Gap, Chemical Composition, Grain Size

Contents

3.1 Introduction

In the present work, five series of doped barium titanate ($BaTiO_3$) ceramic solid solutions with various dopant concentrations were prepared and experimentally characterized using powder X-ray diffraction (PXRD), UV-visible spectrophotometry (UV-vis), scanning electron microscopy (SEM) and energy dispersive X-ray spectroscopy (EDS). Then, by employing the effective analytical techniques namely, Rietveld refinement technique [Rietveld, 1969] and maximum entropy method (MEM) [Collins, 1982], the precise electronic structure, charge density distribution and chemical bonding of the prepared ceramic systems have been investigated. The results obtained from various experimental characterization methods and analytical techniques are presented in chapter 2.

The detailed analysis and discussion of results for the five series of doped $BaTiO_3$ ceramic solid solutions are given in this chapter. The complete details of the sample preparation methods for all the samples are discussed in section 3.2. The structural characterization of all the prepared samples has been carried out using powder X-ray diffraction. All the experimental XRD data sets have been subjected to Rietveld refinement [Rietveld, 1969] through the software JANA 2006 [Petricek et al., 2014]. The complete analysis of experimental XRD patterns, peak shifting trend and the variation of structural parameters due to the influence of various dopants (Sr, Zr, La, Ce and Sr & Zr (co-doping)) with different dopant concentrations are discussed in section 3.3. In addition to this, from the XRD data, the average grain size of all the doped samples has been determined using Scherrer formula [cullity, 2001] and the variation in the average grain size for all the doped sample compositions are discussed in section 3.4.

The optical band gap values have been evaluated from UV-vis absorption analysis (UV-vis) using Tauc plot methodology [Wood and Tauc, 1972]. The variation of optical band gap value with different dopant concentrations is explained in section 3.5.

The morphological and microstructural changes of the various doped titanates have been analyzed using the images obtained from scanning electron microscope (SEM) and are analyzed in section 3.6. The elemental compositions of the synthesized ceramic solid solutions have also been checked using energy dispersive X-ray spectroscopy (EDS) qualitatively and quantitatively. The complete analysis of EDS spectra, the atomic percentages and the weight percentages of the prepared samples are clearly discussed in section 3.7.

The precise electronic structure, chemical bonding between the constituent atoms and charge density distribution in the unit cell of the prepared samples have been completely analyzed by the maximum entropy method (MEM) [Collins, 1982] using the sophisticated softwares PRIMA [Izumi, 2002] and VESTA [Momma, 2008]. The variation in 3D electron density iso-surfaces, 2D electron density contour maps and 1D electron density line profiles due to the effect of various dopants with different dopant concentrations in the crystal structure of barium titanate is discussed in section 3.8.

3.2 Sample preparation

The five differently doped barium titanate ceramic materials were prepared and investigated. The ceramic solid solutions $Ba_{1-x}Sr_xTiO_3$ (BST), $BaTi_{1-x}Zr_xO_3$ (BZT), $BaTi_{1-x}Ce_xO_3$ (BCT) & $Ba_{1-x}Sr_xTi_{0.9}Zr_{0.1}O_3$ (BSZT) were prepared using the conventional high temperature solid state reaction method. The La doped $Ba_{1-x}La_{2x/3}TiO_3$ (BLT) ceramic was prepared using the simple chemical method and sintered at 900 °C for 1 h. The complete experimental details of the solid state reaction method including the calcination temperature, first sintering temperature, second sintering temperature and their corresponding time duration, grinding and regrinding time durations for the four series are given for comparison in table 3.1.

Table 3.1 Comparison of parameters of sample preparation by solid state reaction method.

Samples	Calcination		Grinding Duration (h)	I Sintering		Regrinding Duration (h)	II Sintering	
	Temp. (°C)	Duration(h)		Temp. (°C)	Duration(h)		Temp. (°C)	Duration(h)
BST	1100	10	5	1400	5	-	-	-
BZT	1200	2	3	1450	5	2	1450	5
BCT	1100	6	5	1400	6	5	1400	12
BSZT	1250	15	3	1500	8	5	1500	8

BST - $Ba_{1-x}Sr_xTiO_3$
BZT - $BaTi_{1-x}Zr_xO_3$
BCT - $BaTi_{1-x}Ce_xO_3$
BSZT - $Ba_{1-x}Sr_xTi_{0.9}Zr_{0.1}O_3$

3.3 Structutal analysis of doped titanates

The structural characterization of all five series of doped titanates has been initially performed using powder X-ray diffraction (PXRD). All the XRD data sets have been obtained in the 2θ range of $10°$-$120°$ with the precise step size of $0.02°$. To analyze the structural variation due to the effect of various dopants, all the XRD data sets have been subjected to the Rietveld [Rietveld, 1969] method through the software JANA 2006 [Patricek et al., 2014]. The detailed discussion of experimental XRD profiles and the powder profile refinement is given in this section.

3.3.1 $Ba_{1-x}Sr_xTiO_3$

The structural variation in the crystal structure of barium titanate due to the isovalent substitution of Sr^{2+} (atomic number: 38) at the lattice site of Ba^{2+} (atomic number: 56) has been thoroughly investigated using powder X-ray diffraction patterns.

Figure 2.1(a) shows the raw experimental X-ray diffraction patterns of the synthesized $Ba_{1-x}Sr_xTiO_3$ (x=0.2, 0.4 & 0.6) ceramic samples. The Bragg peaks of the observed experimental XRD patterns are compared with the Joint Committee on Powder Diffraction Standards (JCPDS) data base. All the observed Bragg peaks match very well with the JCPDS data base (PDF # 34-0411). The prepared ceramic samples have been crystallized in the cubic perovskite structure with space group $Pm\bar{3}m$ (space group number: 221). No additional diffraction peaks related to any other secondary phase are detected. When the starting materials are subjected to ball milling, high sintering temperature and lengthy sintering time, the diffusion will be fast which results in a homogeneous product without any secondary phases. So, the high temperature solid state reaction technique guarantees the successful single phase formation of all the synthesized ceramic samples.

Figure 2.1(b) shows the magnified Bragg peaks corresponding to (210) & (211) crystallographic planes which depict the clear peak shift with respect to the incorporation of Sr content. The systematic shift of peaks towards higher 2θ angles by increasing the incorporation of Sr dopant is obviously seen from this figure. This behavior should be attributed to the fact that the substitution of smaller Sr^{2+} ion (ionic radius: 1.13Å) [Shannon, 1976] at the lattice site of Ba^{2+} ion (ionic radius: 1.35 Å) [Shannon, 1976] in the $BaTiO_3$ crystal structure shifted the Bragg peaks towards higher 2θ angles. Moreover, the existence of the well defined and sharp Bragg peaks indicates that the synthesized $Ba_{1-x}Sr_xTiO_3$ ceramic systems have a high degree of crystalline nature.

The structural parameters of BST ceramic samples have been analyzed by the Rietveld refinement method [Rietveld, 1969] through the software JANA 2006 [Patricek et al., 2014]. In the Rietveld [Rietveld, 1969] analysis, by refining various parameters like background, scale factor, shift, lattice constant, profile half width parameters (u, v, w), thermal parameters and occupancy, the difference between experimental and theoretical profiles can be greatly minimized. In this work, the Rietveld refinement [Rietveld, 1969] was done based on BST phase with cubic perovskite structure and space group $Pm\bar{3}m$. The input atomic coordinates (x, y, z) were taken as (0, 0, 0) for barium/strontium, (0.5, 0.5, 0.5) for titanium and (0.5, 0.5, 0) for oxygen from standard Wyckoff position table [Wyckoff, 1963].

The Rietveld fitted [Rietveld, 1969] profiles of BST ceramics for three different Sr concentrations x=0.2, 0.4 & 0.6 are shown in figures 2.2(a), (b) & (c) respectively. In these figures, the cross marks represent the observed powder profiles and continuous lines represent the calculated profiles of the corresponding ceramic systems. The difference between the observed and calculated profiles is shown at the bottom of each figure. The positions of the Bragg peaks are indicated by small vertical lines. The refined profiles show the satisfactory match between the observed and calculated profiles for all the samples. The refined lattice parameters, unit cell volume, density and the reliability indices are given in table 2.1. From this table, it can be seen that, the lattice parameter values are 3.970(14) Å, 3.962(6) Å & 3.941(4) Å for the Sr concentrations x=0.2, 0.4 &

0.6 respectively. The lattice parameter values are found to decrease with the increase of Sr content at the lattice site of Ba. The unit cell volume values also linearly decrease with increasing Sr concentration. This is due to the isovalent substitution of smaller Sr^{2+} ion (ionic radius: 1.13Å) [Shannon, 1976] at the lattice site of larger Ba^{2+} ion (ionic radius: 1.35 Å) [Shannon, 1976]. The values of refined lattice parameters agree well with the previously reported values [Ridha et al., 2009]. The decrement in the cell parameter values with the increasing Sr dopant levels confirms the successful substitution of Sr^{2+} in the $BaTiO_3$ crystal lattice. The values of reliability indices, reliability index for profile (R_P), reliability index for observed structure factors (R_{obs}), and goodness of fit (GOF) indicator confirm the reliable fitting between observed and calculated XRD profiles.

3.3.2 $BaTi_{1-x}Zr_xO_3$

The structural modification in the barium titanate perovskite structure due to the isovalent substitution of Zr^{4+} (atomic number: 40) at the lattice site of Ti^{4+} (atomic number: 22) has been examined in detail using powder X-ray diffraction patterns. Figure 2.3(a) shows the observed XRD patterns of all the Zr doped ceramic samples. In this figure 2.3(a), the XRD patterns indicate that, the synthesized BZT ceramics has the cubic perovskite structure with space group $Pm\bar{3}m$ (space group number: 221), in agreement with the corresponding Joint Committee on Powder Diffraction Standards (JCPDS) data base, PDF # 31-0174. For the Zr composition x=0.04, an additional Bragg peak is observed at 28°, which was indexed as $BaZrO_3$ (PDF # 03-0632). The well-defined sharp Bragg peaks indicate that all the synthesized samples have a high degree of crystalline nature. Figure 2.3(b) shows the enlarged XRD peaks correspond to (110) & (111) crystallographic planes. In this figure, the systematic shift of Bragg peaks towards lower 2θ values with respect to Zr compositions is clearly visualized. This shifting behavior is due to the substitution of the larger Zr^{4+} ion (ionic radius: 0.72Å) [Shannon, 1976] in the place of the smaller Ti^{4+} ion (ionic radius: 0.605Å) [Shannon, 1976].

The experimentally observed XRD profiles of BZT ceramics were subjected to powder profile refinement using the Rietveld method [Rietveld, 1969]. In this analysis, all the profile parameters, structure parameters, corrections and background were refined to get accurate structural information of the prepared samples. The Rietveld refinement [Rietveld, 1969] was performed using the software package JANA 2006 [Petricek et al., 2014].

The positional co-ordinates (x, y, z) were taken as (0, 0, 0) for barium, (0.5, 0.5, 0.5) for titanium/zirconium, and (0.5, 0.5, 0) for oxygen from the Wyckoff position table [Wyckoff, 1963]. The Rietveld [Rietveld, 1969] refined profiles of the BZT ceramics with various Zr concentrations x=0.00, 0.04 & 0.06 are shown in figures 2.4(a), (b) & (c) respectively. The difference between the calculated and observed profiles is shown at the bottom of each figure, which indicates the perfect profile fitting of all the prepared samples. Table 2.2 shows the structural parameters evaluated from the Rietveld refinement [Rietveld, 1969]. The cell parameter values are 4.006(9) Å, 4.017(5) Å & 4.021(4) Å for the Zr doping levels of x=0.00, 0.04 & 0.06 respectively. The cell parameter values increase with increase in Zr content. The unit cell volume values also increase with the Zr doping levels which confirms the lattice expansion by the substitution of larger Zr^{4+} (ionic radius: 0.72Å) [Shannon, 1976] at the lattice site of Ti^{4+} (0.605 Å) [Shannon, 1976]. This lattice expansion behavior also proves that, the Zr^{4+} ions are systematically doped into the $BaTiO_3$ host lattice. The unit cell volume and cell constant values are in good agreement with those reported in the literature [Cheng et al., 2008]. From the values of goodness of fit (GOF) and reliability indices, reliability index for profile (R_P), and reliability index for observed structure factors (R_{obs}) it is again confirmed that, all the powder XRD profiles have been perfectly fitted using the software JANA 2006 [Petricek et al., 2014].

3.3.3 $Ba_{1-x}La_{2x/3}TiO_3$

The variation in the structural parameters due to the aliovalent rare earth substitution of La^{3+} (atomic number: 57) at the lattice site of Ba^{2+} (atomic number: 56) has been analyzed in detail using powder X-ray diffraction patterns.

Figure 2.5(a) shows the experimental raw XRD profiles for BLT ceramics. The Bragg peaks match very well with the reported Joint Committee on Powder Diffraction Standards (JCPDS) data base (PDF # 05-0626). The intensity of the Bragg peaks for the La doped samples is higher than the intensity of the undoped sample. All the prepared samples have been crystallized in tetragonal symmetry with space group $P4mm$ (space group number: 99). In all the prepared ceramic compositions, some extra peaks are also present in addition to the main phase. These additional peaks are identified as $Ba_6Ti_{17}O_{40}$ using the standard diffraction patterns (PDF # 26-0321). The existence of these $Ba_6Ti_{17}O_{40}$ additional phases in the La doped $BaTiO_3$ ceramic systems is reported in the literature too [Wenhu et al., 2009, Urek et al., 2000 & Lin et al., 2002].

Figure 2.5(b) shows the enlarged XRD patterns for (110) & (111) reflections. The Bragg peaks shift towards higher 2θ angles due to the variation in ionic size of Ba^{2+} (ionic radius: 1.35Å) [Shannon, 1976] and La^{3+} (ionic radius: 1.03Å) [Shannon, 1976]. The Bragg peaks corresponding to La composition x=0.015 are slightly shifted towards a lower 2θ angle, probably due to slight lattice disorder.

To analyze the structural parameters and charge density distribution of the synthesized BLT samples, the experimentally observed powder XRD data have been subjected to the

Rietveld refinement [Rietveld, 1969] technique. This powder profile refinement was done through the software package JANA 2006 [Petricek et al., 2014]. The observed profiles were refined by considering the tetragonal structure with *P4mm* space group and one formula unit per primitive cell. The atomic coordinates (x, y, z) were chosen as (0, 0, 0) for barium /lanthanum, (0.5, 0.5, 0.512) for titanium, (0.5, 0.5, 0.023) for O1 and (0.5, 0, 0.486) for O2 from Wyckoff table [Wyckoff, 1963]. The lattice parameters, atomic coordinates, shift, scale factors and other profile related parameters were refined from the observed XRD profiles by comparing them with the theoretically modeled profiles.

The Rietveld [Rietveld, 1969] refined profiles of BLT ceramics are given in figures 2.6(a), (b), (c), (d) & (e) for the La concentrations x=0.000, 0.005, 0.015, 0.020 & 0.025 respectively. The profile fittings confirm the good matching between observed and calculated profiles for both the undoped and doped ceramic compositions. The structural parameters refined from the Rietveld [Rietveld, 1969] method are given in table 2.3. The cell constant values are 4.002(11) Å, 4.000(6) Å, 3.999(3) Å, 3.995(9) Å & 3.994(10) Å for the La concentrations, x=0.000, 0.005, 0.015, 0.020 & 0.025 respectively. The lattice constant values decrease with the incorporation of La. The unit cell volume also decreases with increasing La concentration and the refined lattice constant values are in good agreement with reported literature [Wodecka et al., 2009]. The decrease in unit cell volume and cell parameters is due to the substitution of smaller La^{3+} (ionic radius: 1.03Å) [Shannon, 1976] at the lattice site of larger Ba^{2+} (ionic radius: 1.35Å) [Shannon, 1976]. The phase fractions of main phase (BLT) and additional phase $Ba_6Ti_{17}O_{40}$ are also estimated through Rietveld refinement [Rietveld, 1969]. The amount of impurity phase $Ba_6Ti_{17}O_{40}$ is found to be in the range of 0.7% to 12.5%. The structure factors extracted from the Rietveld refinement [Rietveld, 1969] were further utilized for charge density studies.

3.3.4 $BaTi_{1-x}Ce_xO_3$

The variation of structural parameters due to the isovalent substitution of Ce^{4+} (atomic number: 58) at the lattice site of Ti^{4+} (atomic number: 22) has been examined in detail using powder X-ray diffraction patterns. Figure 2.7(a) shows the raw XRD profiles of $BaTi_{1-x}Ce_xO_3$ (x=0.02, 0.04, 0.06 & 0.08). The Bragg peaks match well with the Joint Committee on Powder Diffraction Standards JCPDS data base (PDF #31-0174). No other additional peaks are detected. This clearly shows the single phase formation of all the Ce doped ceramic samples.

Figure 2.7(b) shows the enlarged Bragg peak corresponding to (220) reflection which clearly explains the shifting behavior of Bragg peaks with respect to the Ce concentrations. The diffraction peaks shift towards lower 2θ angles due to the difference in ionic radii between Ce^{4+} (ionic radius: 0.87Å) [Shannon, 1976] and Ti^{4+} (ionic radius: 0.605Å) [Shannon, 1976]. The systematic shifting of Bragg peaks clearly confirms the successful incorporation of Ce ion at the lattice site of Ti.

The Rietveld [Rietveld, 1969] refinement of the powder X-ray profiles of the prepared ceramic was performed using the software package JANA 2006 [Petricek et al., 2014] by considering the cubic perovskite structure with $Pm\overline{3}m$ space group. The Rietveld refinement [Rietveld, 1969] technique uses the least squares approach to minimize the difference between the experimental profile and theoretically constructed XRD profile. The initial atomic coordinates (x, y, z) were chosen as (0, 0, 0) for barium, (0.5, 0.5, 0.5) for titanium/cerium and (0.5, 0.5, 0) for oxygen from standard Wyckoff position table [Wyckoff, 1963]. Figures 2.8(a), (b), (c) & (d) show the Rietveld [Rietveld, 1969] refined profiles for all the synthesized BCT systems with different Ce concentrations x=0.02, 0.04, 0.06 & 0.08 respectively. At the bottom of each figure, the difference between experimental and theoretical profiles is shown.

The values of lattice constant, unit cell volume and the density with respect to the Ce compositions and reliability indices from the Rietveld refinement [Rietveld, 1969] are given in table 2.4. The lattice constant values are 4.008(8) Å, 4.015(11) Å, 4.030(16) Å & 4.042(18) Å for the Ce doping levels of x=0.02, 0.04, 0.06 & 0.08 respectively. The lattice constant values increase with increasing doping levels of Ce. The unit cell volume values also increase with Ce doping which is due to the substitution of larger Ce^{4+} (ionic radius: 0.87Å) [Shannon, 1976] at the lattice sites of smaller Ti^{4+} (ionic radius: 0.605Å) [Shannon, 1976]. The increment in the lattice constant indicates the lattice expansion effect due to the incorporation of larger Ce at the lattice sites of Ti. The cell parameter values evaluated from the Rietveld [Rietveld, 1969] refinement closely match with the reported values [Chen et al., 2002]. The goodness of fit (GOF) value and reliability indices like, reliability index for observed structure factors (R_{obs}), reliability index for profile (R_P) indicate the better fitting between calculated and observed profiles. The calculated structure factors from Rietveld refinement [Rietveld, 1969] match well with the observed structure factors. The structure factors retrieved from the powder profile refinement were further utilized in charge density analysis.

3.3.5 $Ba_{1-x}Sr_xTi_{0.9}Zr_{0.1}O_3$

The variation in structural parameters due to the isovalent substitution of Sr^{2+} (atomic number: 38) at the lattice site of Ba^{2+} (atomic number: 56) in the Sr & Zr co-doped $BaTiO_3$ systems has been examined in detail using powder X-ray diffraction patterns.

The experimental XRD patterns of BSZT samples with various Sr concentrations x=0.00, 0.05, 0.07 & 0.14 are shown in figure 2.9(a). The observed Bragg peaks match well with the standard pattern JCPDS (PDF # 31-0174) and confirm that the prepared ceramic systems possess cubic perovskite structure with space group $Pm\bar{3}m$ (space group number: 221). Figure 2.9(b) illustrates the enlarged Bragg peaks corresponding to (220) plane which clearly show that the peaks shift towards the higher 2θ angle with the increased doping levels of Sr. This peak shifting trend is due to the dopant of smaller Sr^{2+} (ionic radius: 1.13Å) in place of larger Ba^{2+}(ionic radius: 1.35Å) [Shannon, 1976] [Shannon, 1976]. This peak shift qualitatively confirms that, the Sr dopant is systematically dissolved in the chosen compositional range.

The structural studies of the prepared BSZT samples were carried out by Rietveld refinement [Rietveld, 1969] using powder XRD data. The lattice parameters along with the correction parameters were also included in the refinement using the software JANA 2006 [Petricek et al., 2014]. In the synthesized series of samples, with ABO_3 type perovskite structure, Ba & Sr ions occupy A-sites and Ti & Zr ions occupy B-sites. The experimental XRD profiles of BSZT were refined by considering the cubic setting with one formula unit per primitive cell. The Ba & Sr atoms are fixed at (0, 0, 0) and Ti & Zr atoms are fixed at (0.5, 0.5, 0.5) and oxygen is at (0.5, 0.5, 0) positions as given in Wyckoff position table [Wyckoff, 1963]. The fitted profiles of $Ba_{1-x}Sr_xTi_{0.9}Zr_{0.1}O_3$ (x=0.00, 0.05, 0.07, & 0.14) are given in figures 2.10(a), (b), (c) & (d) respectively. In these figures, the cross marks show the observed profile, the continuous lines indicate the calculated profile and the positions of the Bragg peaks are denoted by the vertical lines. The difference between the observed and calculated profiles is given at the bottom of each figure. The fitted profiles reveal the good matching between the observed and calculated profiles for all the four compositions. The refined structural parameters, reliability index for profile (R_P), reliability index for observed structure factors (R_{obs}) and Goodness of fit (GOF) indicator values are given in table 2.5. The cell constant values are 4.031(8) Å, 4.025(9) Å, 4.012(2) Å & 4.005(6) Å for the Sr compositions x=0.00, 0.05, 0.07 & 0.14 respectively. The cell constant and unit cell volume decrease with increasing incorporation of Sr. The decrease in cell constant and the shrinkage in unit cell volume with Sr concentrations are due to the substitution of smaller Sr^{2+} (ionic radius: 1.13Å) [Shannon, 1976] at the lattice site of Ba^{2+} (ionic radius: 1.35Å) [Shannon, 1976]. The evaluated lattice parameters are in agreement with the reported values [Thongtha et al., 2014].

The refined structural parameters for all the doped titanates are given in table 3.2 for comparison. When smaller Sr^{2+} is substituted at the lattice site of Ba^{2+}, the cell parameter and the unit cell volume decrease with increase in Sr concentration. When larger Zr^{4+} is substituted at the lattice site of Ti^{4+}, the cell parameter and the unit cell volume increase with Zr concentration. With incorporation of smaller La^{3+} at the lattice site of Ba^{2+}, the cell parameter and the unit cell volume decrease with La concentration. When larger Ce^{4+} is substituted at the lattice site of Ti^{4+}, the cell parameter and the unit cell volume increase with Ce concentration. In the Sr & Zr co-doped ceramic systems, by the incorporation of smaller Sr^{2+} in the lattice site of Ba^{2+} the cell parameter and the unit cell volume decrease with Sr concentration. The variation in cell constant and unit cell volume is due to the difference in ionic radius between the dopant ion and the host ion.

Table 3.2 *Comparison of refined structural parameters for doped titanates.*

Samples	Crystall ographi c system	Conc. (x)	Lattice parameters			
			a (Å)	b (Å)	c (Å)	Unit cell volume (Å³)
BST	Cubic $Pm\bar{3}m$	0.20	3.970(14)	3.970(14)	3.970(14)	62.60(2)
		0.40	3.962(6)	3.962(6)	3.962(6)	62.22(9)
		0.60	3.941(4)	3.941(4)	3.941(4)	62.23(6)
BZT	Cubic $Pm\bar{3}m$	0.00	4.006 (9)	4.006(9)	4.006(9)	64.32(5)
		0.04	4.017(5)	4.017(5)	4.017(5)	64.86(1)
		0.06	4.021(4)	4.021(4)	4.021(4)	65.02(3)
BLT	Tetra -gonal $P4mm$	0.000	4.002(11)	4.002(11)	4.017(14)	64.35(32)
		0.005	4.000(6)	4.000(6)	4.021(8)	64.37(24)
		0.015	3.999(3)	3.999(3)	4.021(2)	64.32(12)
		0.020	3.995(9)	3.995(9)	4.017(11)	63.12(9)
		0.025	3.994(10)	3.994(10)	4.016(2)	63.08(17)
BCT	Cubic $Pm\bar{3}m$	0.02	4.008(8)	4.008(8)	4.008(8)	64.42(1)
		0.04	4.015(11)	4.015(11)	4.015(11)	64.72(1)
		0.06	4.030(16)	4.030(16)	4.030(16)	65.52(2)
		0.08	4.042(18)	4.042(18)	4.042(18)	66.07(3)
BSZT	Cubic $Pm\bar{3}m$	0.00	4.031(8)	4.031(8)	4.031(8)	65.43(1)
		0.05	4.025(9)	4.025(9)	4.025(9)	65.22(1)
		0.07	4.012(2)	4.012(2)	4.012(2)	64.57(2)
		0.14	4.005(6)	4.005(6)	4.005(6)	64.12(3)

BST - $Ba_{1-x}Sr_xTiO_3$
BZT - $BaTi_{1-x}Zr_xO_3$
BLT - $Ba_{1-x}La_{2x/3}TiO_3$
BCT - $BaTi_{1-x}Ce_xO_3$
BSZT - $Ba_{1-x}Sr_xTi_{0.9}Zr_{0.1}O_3$
Conc. – concentration

3.4 Grain size analysis of doped titanates

The average grain size of all the doped titanates has been evaluated using Scherrer formula [Cullity, 2001], $t = 0.9\lambda/\beta cos\theta$, where, λ is the wavelength of X-rays, 1.54056 Å, β is the full width at half maximum (FWHM) in radians, and θ is the Bragg angle, t is the average grain size, which is the average size of coherently diffracting domains. The evaluation of grain size has been done by using the full width at half maximum (FWHM) of the observed X-ray diffraction peaks and observed Bragg angles through the GRAIN software [Saravanan, personal communication].

The average grain size has been evaluated for $Ba_{1-x}Sr_xTiO_3$ samples with different Sr doping levels for, x=0.2, 0.4 & 0.6 and found to be in the range of 22 nm to 36 nm and the values are in agreement with the previously reported values [Balachandran et al., 2008].

The average grain size of $BaTi_{1-x}Zr_xO_3$ samples with different Zr doping levels for, x=0.00, 0.04 & 0.06 has been evaluated and found to be in the range of 23 nm to 28 nm which is in good agreement with the previous studies [Dash et al., 2014].

The average grain size corresponding to $Ba_{1-x}La_{2x/3}TiO_3$ samples with various La concentrations for, x=0.000, 0.005, 0.015, 0.020 & 0.025 has been evaluated and found to be in the range of 21 nm to 36 nm. The evaluated values closely match with the earlier reports [Vijatovic et al., 2010].

The average grain size of the $BaTi_{1-x}Ce_xO_3$ ceramics with various Ce doping levels for, x=0.02, 0.04, 0.06 & 0.08 has been evaluated and found to be in the range of 17 nm to 26 nm.

The average grain size of the $Ba_{1-x}Sr_xTi_{0.9}Zr_{0.1}O_3$ (Sr & Zr (co-doping)) ceramics with various Sr doping levels for, x=0.00, 0.05, 0.07 & 0.14 has been evaluated and found to be in the range of 21 nm to 32 nm.

Table 3.3 gives the comparison of average grain size range for doped titanates from XRD patterns of doped titanates. In a series, for each composition, the average grain size has been evaluated and the range of average grain size for all compositions for a series is given. The powder XRD gives only the size of the coherently diffracting domains.

Table 3.3 Comparison of average grain size range for doped titanates from XRD.

Samples	Average grain size range (nm)
BST	22 - 36
BZT	23 - 28
BLT	21 - 36
BCT	17 - 26
BSZT	21 - 32

BST - $Ba_{1-x}Sr_xTiO_3$
BZT - $BaTi_{1-x}Zr_xO_3$
BLT - $Ba_{1-x}La_{2x/3}TiO_3$
BCT - $BaTi_{1-x}Ce_xO_3$
BSZT - $Ba_{1-x}Sr_xTi_{0.9}Zr_{0.1}O_3$

In the present work, for all the doped titanates, the average grain size is obtained in the nanometer range. This is due to the fact that, after completing the preparation, sintered pellets were forcefully ground using an agate mortar and the resulting fine powder samples were sent for characterization studies. So the grain size has been reduced to nanometer scale.

3.5   Optical analysis of doped titanates

The optical band gap of all the doped $BaTiO_3$ ceramic solid solutions has been evaluated using the absorption data obtained using a UV-vis spectrophotometer. From the UV-vis absorption spectra which are obtained in the range of 200 nm to 2000 nm, the optical band gap has been evaluated through Tauc plot [Wood and Tauc, 1972] methodology which is explained in chapter 1, section 1.11.3.

3.5.1 $Ba_{1-x}Sr_xTiO_3$

The UV-vis absorption spectra of the $Ba_{1-x}Sr_xTiO_3$ samples doped with various Sr concentrations (x=0.2, 0.4 & 0.6) are shown in figure 2.11. The Tauc plot [Wood and Tauc, 1972] for BST samples for different Sr compositions is given in figure 2.12. The optical band gap values are given in table 2.6. The optical band gap values are 3.128 eV, 3.098 eV & 3.067 eV for the Sr compositions, x=0.2, 0.4 & 0.6 respectively. The band gap values closely agree with the previously reported studies [Souza et al., 2006]. In this study, optical band gap decreases with the increasing doping levels of Sr. The decrement in the band gap values implies the reduction in insulating property of the Sr doped $BaTiO_3$ ceramic system in the selected compositional range.

3.5.2 $BaTi_{1-x}Zr_xO_3$

The UV-vis absorption spectra corresponding to the $BaTi_{1-x}Zr_xO_3$ samples with different doping levels of Zr, for x=0.00, 0.04 & 0.06 are depicted in figure 2.13. The Tauc plot is given in figure 2.14. The optical band gap values are given in table 2.7. For the undoped $BaTiO_3$ sample, the band gap value is 3.181 eV. For Zr concentrations x=0.04 and x=0.06, the band gap values are 3.087 eV and 3.043 eV respectively. In this study, optical band gap energy decreases with the increasing doping levels of Zr. The obtained values are in good agreement with the previously reported values for Zr doped $BaTiO_3$ ceramic systems [Cavalcante et al., 2008].

3.5.3 $Ba_{1-x}La_{2x/3}TiO_3$

The UV-vis absorption spectra for $Ba_{1-x}La_{2x/3}TiO_3$ samples with various La concentrations of x=0.000, 0.005, 0.015, 0.020 & 0.025 are shown in figure 2.15. The Tauc plot is given in figure 2.16 and the optical band gap values are given in table 2.8. The optical band gap value for the undoped $BaTiO_3$ sample is 3.146 eV. For the La doped samples x=0.005, 0.015, 0.020 & 0.025 the band gap values are 3.135 eV, 3.156 eV, 3.146 eV & 3.147 eV respectively. Since the differences in La doping levels are minimal, a very small variation in the band gap values is observed. La substituted samples are known to be associated with the structural disorder due to the creation of A-site vacancies and distortion in TiO_6 clusters. The existence of structural defects can form the localized states between the conduction band and valance band which results in the small variation in band gap (E_g) values [Ganguly et al., 2013].

3.5.4 $BaTi_{1-x}Ce_xO_3$

$BaTi_{1-x}Ce_xO_3$ samples with various concentrations of Ce (x=0.02, 0.04, 0.06 & 0.08) are subjected to UV-vis analysis for band gap evaluation. The UV-vis absorption spectra are shown in figure 2.17. The Tauc plot of BCT series is given in figure 2.18. The optical band gap values are given in table 2.9. The optical band gap values for Ce concentrations x=0.02, 0.04, 0.06 & 0.08 are 3.165 eV, 3.154 eV, 3.136 eV & 3.185 eV respectively. The optical band gap values are found to decrease with increasing doping levels of Ce which fact clearly explains that, the substitution of Ce in the $BaTiO_3$ crystal lattice reduces the insulating property of the system.

3.5.5 $Ba_{1-x}Sr_xTi_{0.9}Zr_{0.1}O_3$

$Ba_{1-x}Sr_xTi_{0.9}Zr_{0.1}O_3$ (Sr & Zr (co-doping) samples with various Sr concentrations (x=0.00, 0.05, 0.07 & 0.14) are subjected to UV-vis analysis for the evaluation of the band gap value. The UV-vis absorption spectra for various Sr concentrations are shown in figure

2.19. The Tauc plot for the BSZT samples is given in figure 2.20. The optical band gap values are given in table 2.10. For undoped sample, the band gap value is 3.208 eV. For the Sr dopant levels x=0.05, 0.07 & 0.14, the band gap values are 3.193 eV, 3.174 eV & 3.183 eV respectively. The optical band gap values decrease with the increasing incorporation of Sr content which fact indicates the slight reduction in the insulating property of the BSZT systems.

Table 3.4 explains the comparison of optical band gap ranges for doped titanates from the UV-vis analysis. The optical band gap values for the $Ba_{1-x}Sr_xTiO_3$, $BaTi_{1-x}Zr_xO_3$, $BaTi_{1-x}Ce_xO_3$ & $Ba_{1-x}Sr_xTi_{0.9}Zr_{0.1}O_3$ ceramic solid solutions decrease with the increase in dopant concentrations. The decrease in the optical band gap indicates the slight reduction of the insulating property of the above mentioned four doped titanates. But, La doped $BaTiO_3$ prepared according to the formula $Ba_{1-x}La_{2x/3}TiO_3$, is charge compensated by A-site cation vacancies (V_{Ba}) [Morrison et al., 2001 & Chen et al., 2011]. La doped $BaTiO_3$ charge compensated in this manner should remain insulating due to the immobility of cation vacancies. So, a systematic variation of the band gap was not observed in this La doped $Ba_{1-x}La_{2x/3}TiO_3$ system.

Table 3.4 *Comparison of optical band gaps for doped titanates from the UV-vis analysis.*

Samples	Optical band gap (eV)		
BST	3.067	-	3.128
BZT	3.043	-	3.181
BLT	3.135	-	3.156
BCT	3.136	-	3.185
BSZT	3.174	-	3.208

BST - $Ba_{1-x}Sr_xTiO_3$
BZT - $BaTi_{1-x}Zr_xO_3$
BLT - $Ba_{1-x}La_{2x/3}TiO_3$
BCT - $BaTi_{1-x}Ce_xO_3$
BSZT - $Ba_{1-x}Sr_xTi_{0.9}Zr_{0.1}O_3$

3.6 Morphological analysis of doped titanates

The surface morphology of all the doped titanate compositions has been investigated using scanning electron microscopy (SEM) by obtaining images with various magnifications ranging from ×5000 to ×25000. The average particle size has also been evaluated from SEM images. The particle size evaluated from the SEM image is not directly comparable with the grain size from XRD.

3.6.1 $Ba_{1-x}Sr_xTiO_3$

Figures 2.21(a), (b) & (c) show the scanning electron microscope (SEM) images corresponding to ×10000 magnification for the Sr concentrations x=0.2, 0.4 & 0.6 respectively. It is observed that, the incorporation of strontium ion at the barium lattice site changes the microstructure and surface morphology of the prepared samples significantly. For all the Sr concentrations, the particles are irregular in shape and size. In the series of samples, for each composition, there is an average particle size determined from the SEM images. The range of the average particle size for all the Sr compositions is 1 μm to 2 μm which is in close agreement with the previously reported values [Suasmoro et al., 2000].

3.6.2 $BaTi_{1-x}Zr_xO_3$

Figures 2.22(a), (b) & (c) show the scanning electron microscope (SEM) images of the BZT samples with various doping levels (x=0.00, 0.04 & 0.06) corresponding to a magnification level ×25000. Particles with irregular sizes are seen from the SEM images. With the incorporation of Zr in the crystal structure of $BaTiO_3$, the grains are well developed and the domains are clearly visible. The average particle size is in the range of 1 μm to 1.7 μm. The presence of well developed larger grains with the irregular sizes may be due to the variation in the kinetics of movement from one boundary to another [Badapanda et al., 2011].

3.6.3 $Ba_{1-x}La_{2x/3}TiO_3$

Figures 2.23(a), (b), (c), (d) & (e) show the scanning electron microscope (SEM) images of BLT samples with different La doping levels (x=0.000, 0.005, 0.015, 0.020 & 0.025) with a magnification level of ×10000. All the samples possess dense microstructure consisting of small grains with nearly spherical shape. The fine-grained microstructure with narrow sized particle distribution is evident for the undoped and low-level lanthanum doped samples. The microstructure of the samples also consists of large voids, which are also clearly visible from SEM micrographs. The surface morphology of the particles for all the compositions is uniform in shape due to the chemical method of preparation in which the growth will be natural and force-free. The average particle size is determined from SEM images and it is in the range of 0.2 μm to 0.4 μm which is in close agreement with the earlier studies [Ramoska et al., 2010].

3.6.4 $BaTi_{1-x}Ce_xO_3$

Figures 2.24(a), (b), (c) & (d) show the scanning electron microscope (SEM) images of the prepared BCT samples with various Ce doping levels (x=0.02, 0.04, 0.06 & 0.08) for a magnification level ×10000. The incorporation of Ce in the crystal structure of $BaTiO_3$ leads to the dense microstructure. Formation of aggregated particles with low porosity has been clearly seen from all the Ce doped samples. The average particle size is determined from SEM images and it is in the range of 0.9 μm to 1.4 μm

3.6.5 $Ba_{1-x}Sr_xTi_{0.9}Zr_{0.1}O_3$

Figures 2.25(a), (b), (c) & (d) show the SEM images corresponding to ×15000 magnification for different Sr doping levels. The formation of particles with well developed grain boundaries shows that, the simultaneous substitution of Sr and Zr into the $BaTiO_3$ lattice promotes good grain growth. So, large sized and well developed particles are seen from the images. The average particle size is determined from SEM images and it is in the range of 5 μm to 9 μm.

Table 3.5 *Comparison of average particle size range and shape for doped titanates from SEM images.*

Samples	Shape of the particle	Average particle size range (μm)
BST	irregular	1.0-2.0
BZT	irregular	1.0-1.7
BLT	narrow size, uniform nearly spherical with large voids	0.2-0.4
BCT	aggregated with low porosity	0.9-1.4
BSZT	large sized and well developed particles	5.0-9.0

BST - $Ba_{1-x}Sr_xTiO_3$
BZT - $BaTi_{1-x}Zr_xO_3$
BLT - $Ba_{1-x}La_{2x/3}TiO_3$
BCT - $BaTi_{1-x}Ce_xO_3$
BSZT - $Ba_{1-x}Sr_xTi_{0.9}Zr_{0.1}O_3$

Table 3.5 gives the comparison of average particle size ranges and shape for doped titanates from SEM images. In a series, each composition has an average particle size. The range of average particle sizes for all the doped titanates are given in table 3.5. The average particle size evaluated from the SEM micrographs is in the micrometer scale. Smaller particle size (0.2 μm - 0.4 μm) is obtained for lanthanum doped barium titanate ceramic (BLT) samples prepared through the chemical method. The shape of the particles is also uniform in the naturally grown samples (chemical method). The particle size is larger (5 μm - 9 μm) for Sr & Zr co-doped BaTiO$_3$ ceramic (BSZT) solid solutions prepared through the solid state reaction method. This proves that, the BSZT samples prepared through the high temperature solid state reaction technique promotes good grain growth.

3.7 Elemental analysis of doped titanates

The chemical compositions of all the prepared samples have been investigated by characterizing the samples using energy dispersive X-ray spectroscopy (EDS). The EDS spectra qualitatively verify the chemical compositions of the prepared systems. The atomic and mass percentages of the doped titanates have also been verified by EDS quantitative analysis.

3.7.1 Ba$_{1-x}$Sr$_x$TiO$_3$

Figures 2.26(a), (b) & (c) show the energy dispersive X-ray spectroscopy (EDS) spectra of the BST samples for different Sr doping concentrations, x=0.2, 0.4 & 0.6 respectively. The presence of the X-ray peaks corresponding to various atoms shows the successful substitution of Sr in the Ba$_{1-x}$Sr$_x$TiO$_3$ solid solutions. The elemental compositions in terms of mass percentages and atomic percentages are listed in table 2.11. The atomic percentages and mass percentages of the prepared samples confirm the stoichiometry of the prepared samples.

3.7.2 BaTi$_{1-x}$Zr$_x$O$_3$

Figures 2.27(a), (b) & (c) represent the energy dispersive X-ray spectroscopy (EDS) spectra of all the synthesized BZT samples for different Zr doping concentrations x=0.00, 0.04 & 0.06 respectively. From this spectral analysis, it is evident that, all the prepared BZT ceramics are composed only of Ba, Ti, Zr and O atoms. The atomic percentages and weight percentages of the constituent elements in the samples are given in table 2.12, which confirm the stoichiometry of all the synthesized BZT samples.

3.7.3 Ba$_{1-x}$La$_{2x/3}$TiO$_3$

The EDS spectra of the prepared samples are shown in figures 2.28(a), (b), (c), (d) & (e) for the La doping levels, x=0.000, 0.005, 0.015, 0.020 & 0.025 respectively. The elemental compositions of the synthesized samples in terms of mass percentages and atomic percentages corresponding to Ba, Ti and O atoms are given in table 2.13. The characteristic X-ray peak of each element has different energy values. The X-ray energies corresponding to La and Ti are almost the same (La of L is 4.525 KeV, Ti of K is 4.508 KeV), and hence of the X-ray peaks of La & Ti overlap. [Mancic et al., 2008]

3.7.4 $BaTi_{1-x}Ce_xO_3$

Figures 2.29 (a), (b), (c) & (d) show the EDS spectra of all the Ce doped samples with doping levels x=0.02, 0.04, 0.06 & 0.08 respectively. The characteristic X-ray peaks of each element in the spectrum confirm the presence of Ba, Ti, Ce and O atoms in the prepared samples. No additional peak has been detected in all the spectra which confirm the phase purity of the prepared Ce doped samples. The atomic percentages and mass percentages of the prepared samples are given in table 2.14. This quantitative EDS analysis confirms the conservation of stoichiometric composition of all the Ce doped samples.

3.7.5 $Ba_{1-x}Sr_xTi_{0.9}Zr_{0.1}O_3$

The EDS spectra of all the synthesized samples are shown in figures 2.30(a), (b), (c) & (d). The presence of characteristic X-ray peaks corresponding to the main constituent elements of the prepared BSZT samples Ba, Ti, Zr, Sr and O are seen in figures 2.30(a), (b), (c) & (d). No extra peak corresponding to any other impurity elements has been detected from these EDS spectra for all the compositional range and this confirms the single phase formation of all the synthesized BSZT samples.

3.8 Charge density analysis of doped titanates

The precise electronic structure, chemical bonding and electron density analysis of all the five series of doped titanates were effectively analyzed using the maximum entropy method (MEM) [Collins, 1982]. The structure factors extracted from the Rietveld refinement [Rietveld, 1969] were utilized for MEM [Collins, 1982] refinement. The MEM [Collins, 1982] calculations have been carried out by employing the software PRIMA [Izumi, 2002] by considering 64×64×64 pixels along the a, b and c axes of the unit cell. The resultant three dimensional and two dimensional electron density distribution are plotted with the help of the visualization software VESTA [Momma, 2008].

3.8.1 $Ba_{1-x}Sr_xTiO_3$

Figures 2.31 (a), (b) & (c) show the 3D electron density distribution of $Ba_{1-x}Sr_xTiO_3$ for Sr doping concentrations, x=0.2, 0.4 and 0.6, constructed with the similar iso-surface level of 0.8 e/Å³. The positions of Ba, Ti and O atoms and the electron density distribution around them are clearly seen in these figures 2.31 (a), (b) & (c). Ba atoms are situated at the corners, O atoms are at the face centered positions and the Ti atom is at the body centered position of the 3D unit cell which shows the perfect perovskite structure. Due to the substitution of Sr at the lattice site of Ba, the electron density distribution around the Ba, Ti & O atoms has been considerably changed. The three dimensional view of enlarged portion of Ba-O and Ti-O bonds for three different Sr doping levels is given in figure 2.32. The increasing incorporation of Sr in the crystal structure of BaTiO₃ increases the volume of the electron density distribution around Ba atom. The reduction in the bond lengths of Ba-O and Ti-O has also been clearly visualized from this figure.

To analyze the nature of bonding qualitatively, 2D electron density contour maps have been constructed for the Sr doped samples for three different crystallographic planes, with contour interval of 0.04 e/Å³. Figure 2.33(a) shows the 3D unit cell of BST with (100) plane shaded and (b), (c) & (d) represent 2D electron density contour maps of the samples for three different Sr concentrations x=0.2, 0.4 & 0.6 respectively with enlarged view of bonding between the Ba and O atoms on the crystallographic plane (100). The electron density contour lines around the Ba atom are elongated towards the face centered oxygen atoms. This elongation is due to the electronegativity difference between oxygen (electronegativity: 3.44) and barium (electronegativity: 0.89). The atom with greater electronegativity has the tendency of attracting the electrons towards it, like the oxygen atom, in this case. The elongation of electron density contours towards oxygen atom increases with the increasing incorporation of Sr content. The distribution of contour lines between Ba and O atoms explains the ionic nature of Ba-O bond. To examine the nature of Ti-O bond in detail, 2D electron density contour maps are drawn for two different crystallographic planes, (200) and (101).

Figure 2.34(a) shows the 3D unit cell of BST with (200) plane shaded and (b), (c) & (d) represent the 2D electron density contour maps of the samples for the Sr concentrations x=0.2, 0.4 & 0.6 respectively with enlarged view of bonding between the Ti and O atoms. Figure 2.35(a) shows the 3D unit cell of BST with the (101) plane shaded and (b), (c) & (d) represent the 2D electron density contour maps of the samples for the Sr concentrations, x=0.2, 0.4 & 0.6 respectively with enlarged view of bonding between Ti and O atoms. The density of the contour lines along the Ti-O bonding region increases with the increasing doping levels of Sr. The overlapping of contour lines along the Ti-O bond exhibits the covalent nature of bonding which is due to the result of hybridization between the O-2p and Ti-3d orbitals. When the smaller Sr ion is substituted at the lattice site of larger Ba, the volume of the unit cell become smaller and Ti ions can be in good touch with all the O ions which increases the covalent nature of the Ti-O bond.

The quantitative analysis of chemical bonding between the constituent atoms has been carried out by drawing the 1D electron density line profiles of Ba-O and Ti-O bonds. Figure 2.36 shows the 1D electron density line profiles for the Ba-O bond for three different Sr concentrations. Figure 2.37 shows the 1D electron density line profiles for the Ti-O bond for three different Sr concentrations.

The bond lengths and mid bond electron density values of Ba-O and Ti-O bonds of $Ba_{1-x}Sr_xTiO_3$ (x=0.2, 0.4 & 0.6) are given in table 2.15. The value of mid bond electron density along the Ba-O bond decreases with respect to Sr incorporation. The values of electron density show the increase in ionic nature of the prepared samples due to Sr doping. Along the Ti-O bond, the mid bond density values show increasing trend with the addition of Sr content which confirms the decrement in ionic nature between Ti and O ions. The bond length of the Ba-O bond decreases from 2.807 Å to 2.787 Å. Also, the bond length of the Ti-O bond decreases from 1.985 Å to 1.971 Å. The bond lengths of Ba-O and Ti-O are found to decrease with respect to the Sr doping levels, which is due to the lattice shrinkage, as confirmed by XRD results.

3.8.2 $BaTi_{1-x}Zr_xO_3$

The charge density mapping and bonding interaction between the constituent atoms in Zr doped $BaTi_{1-x}Zr_xO_3$ ceramic systems for various concentrations x=0.00, 0.04 & 0.06 have been analyzed.

Figures 2.38(a), (b) & (c) show the 3D electron density distribution in the unit cell of the prepared systems. In these figures 2.38(a), (b) & (c), the positions of Ba, Ti and O atoms with their electron density distribution around them are clearly visualized. The electron density mapping and inter-atomic bonding between the atoms have been analyzed qualitatively by constructing 2D electron density contour maps for the two most significant crystallographic planes (100) and (200). In the (100) plane, Ba atoms are at the corners and O atoms are at the face center. In the (200) plane, Ti atom is at the center and the O atoms are at the face centers. By analyzing the electron density distribution on (100) and (200) planes, the nature of Ba-O and Ti-O bonds can be clearly elucidated.

Figure 2.39(a) represents the 3D unit cell with the (100) plane shaded. Figures 2.39(b), (c) & (d) explain the 2D electron density distribution with the enlarged view of the Ba-O bond on the (100) plane for various Zr concentrations, x=0.00, 0.04 &0.06 respectively. In figure 2.39(b), it is observed that, there is not much charge linkage between the Ba and O atoms which clearly indicates the ionic nature of the Ba-O bond. Substitution of Zr (atomic number: 40) at the lattice site of Ti (atomic number: 22) affects the charge density around the neighboring Ba and O atoms in the $BaTiO_3$ crystal structure and hence the ionic nature reduces slightly, which is clearly visualized in figures 2.39(c) & (d). The chemical bonding between the Ti and O atoms has been analyzed by examining the electron density distribution around the atoms on the (200) plane. Figure 2.40(a) shows the 3D unit cell with the (200) plane shaded, and figures 2.40(b), (c) & (d) depict the 2D electron density distributions with an enlarged view of the Ba-O bond on the (200) plane for three various Zr concentrations, x=0.00, 0.04 & 0.06. The electron density contours along Ti and O increase with the increasing doping levels of Zr. The overlapping of electron density along the Ti-O bond path indicates the covalent nature of bonding between the Ti and O atoms. This behavior is attributed to the large overlapping of the Ti-3d orbitals and O-2p orbitals [Piskunov et al., 2004].

Quantitative analysis of the nature of chemical bonding has been carried out by drawing 1D electron density line profiles along the Ba-O and Ti-O bonds. Figure 2.41 shows the 1D electron density line profiles along the Ba-O bond and figure 2.42 shows the 1D line profiles of the Ti-O bond for the selected Zr concentrations. The accurate mid bond electron density values and the variation of bond lengths with respect to the Zr concentrations are presented in table 2.16. The Ba-O bond length value increases from 2.833 Å to 2.843 Å. Similarly, the bond length value of the Ti-O bond increases from 2.003 Å to 2.010 Å. The increment in bond lengths are due to the incorporation of larger Zr^{4+} (ionic radius: 0.72 Å) [Shannon, 1976] in the place of smaller Ti^{4+} (ionic radius: 0.605 Å) [Shannon, 1976] and also confirms the lattice expansion effect as confirmed by XRD results. For the undoped sample, the electron density value of the Ba-O bond at the bond critical point (BCP) is 0.291 $e/Å^3$. The electron density value increase up to 0.401 $e/Å^3$ with increasing Zr concentration, which shows the weakening of ionic nature along the Ba-O bond. The electron density value at the bond critical point (BCP) for the undoped sample is 0.657 $e/Å^3$, whereas this value increases up to 0.738 $e/Å^3$ for the increased doping levels of Zr due to more electron interaction which confirms the enhancement of covalent nature along the Ti-O bond.

3.8.3 $Ba_{1-x}La_{2x/3}TiO_3$

The bonding nature and electron density distribution of La doped $Ba_{1-x}La_{2x/3}TiO_3$ ceramic systems with various concentrations x=0.000, 0.005, 0.015, 0.020 & 0.025 have been accurately elucidated by the maximum entropy method (MEM) [Collins, 1982].

Figures 2.43(a), (b), (c), (d) & (e) show the 3D view of the unit cell for various La concentrations which are constructed by considering the same iso-surface level of 1 $e/Å^3$. The positions of barium, titanium and oxygen atoms and the electron density distribution around the atoms have been clearly seen from these figures 2.43(a), (b), (c), (d) & (e). The accurate electron density distribution for two different crystallographic planes (001) and (002) has been analyzed by drawing the 2D electron density contour maps. Figure

2.44(a) shows the 3D unit cell with the (001) plane shaded. Figures 2.44(b), (c), (d), (e) & (f) represent the 2D electron density contour maps with the enlarged view of the Ba-O1 bond on (001) plane for five different La concentrations x=0.000, 0.005, 0.015, 0.020 & 0.025 respectively. 2D electron density contour maps for all the compositions clearly indicate the predominant ionic nature between Ba and O1.

Figure 2.45(a) shows the 3D unit cell with the (002) plane shaded, figures 2.45(b), (c), (d), (e) & (f) show the 2D electron density contour maps with the enlarged view of the Ti-O2 bond for five different La concentrations x=0.000, 0.005, 0.015, 0.020 & 0.025 respectively. The two dimensional electron density contour maps show the explicit charge sharing between the titanium and oxygen atoms which explain the partial covalent nature of the samples. The covalent nature arises due to the hybridization of Ti-3d and O-2p orbitals.

Accurate electron density variation along Ba-O1 and Ti-O2 due to La doping has been analyzed quantitatively by drawing the 1D line profiles. Numerical values of electron density are extracted from the 1D line profiles. Figure 2.46 shows the 1D electron density line profile of the Ba-O1 bond and figure 2.47 shows the 1D electron density line profile of the Ti-O2 bond for various La concentrations. The bond length variation and mid bond density distribution between the constituent atoms are given in table 2.17. The Ba-O1 and Ti-O2 bond lengths decrease with the increasing incorporation of La, due to the substitution of smaller La (ionic size: 1.03 Å) [Shannon, 1976] in the place of larger Ba (ionic size: 1.35 Å) [Shannon, 1976]. The mid bond electron density value between Ba and O1 for the undoped sample is 0.309 e/Å^3 and for the La doped samples, the mid bond density values are 0.278 e/Å^3, 0.283 e/Å^3, 0.293 e/Å^3 and 0.288 e/Å^3. The values of mid bond density confirm the existence of predominant ionic nature between barium and oxygen. The mid bond electron density along yhe Ti-O2 bond corresponding to the undoped sample is 0.674 e/Å^3. For the La doped samples, with the concentrations x=0.005, 0.015, 0.020 & 0.025, the mid bond electron density values are evaluated as 0.649 e/Å^3, 0.659 e/Å^3, 0.742 e/Å^3 & 0.724 e/Å^3 respectively. These values of mid bond electron densities confirm the covalent nature of the Ti-O2 bond. There is no significant change in the electron density values for different La concentrations.

From the defect chemistry analysis, La^{3+} substitution at the lattice site of Ba^{2+} (A-site of BaTiO$_3$ structure) creates the charge imbalance which should be compensated by three different mechanisms: Compensation may either by A-site cation vacancies (V$_{Ba}$) or B-site cation vacancies (V$_{Ti}$) or by electrons [Morrison et al., 2001]. It was explained that, the La doped BaTiO$_3$ solid solutions prepared with formula Ba$_{1-x}$La$_x$Ti$_{1-x/4}$O$_3$ create B-site cation vacancy (V$_{Ti}$). The solid solutions with formula Ba$_{1-x}$La$_x$TiO$_3$ give additional electrons to the system. In this present work, La substituted BaTiO$_3$ solid solutions synthesized according to the formula $Ba_{1-x}La_{2x/3}TiO_3$ are charge compensated by A-site cation vacancies (V_{Ba}). The number of cation vacancy increases with the La doping levels [Ganguly et al., 2013]. The mid bond electron density values do not change significantly with La doping due to the presence of cation vacancies. La doped $BaTiO_3$ ceramic compositions charge compensated in this way should remain insulating due to the immobility of cation vacancies. The bond lengths of Ba-O1 and Ti-O2 decrease with La addition which is due to the lattice shrinkage consistent with the XRD results.

3.8.4 $BaTi_{1-x}Ce_xO_3$

The electronic structure of Ce doped $BaTiO_3$ and the distribution of charges around the Ba-O and Ti-O bonds in the synthesized $BaTi_{1-x}Ce_xO_3$ (x=0.02, 0.04, 0.06 & 0.08) ceramic samples have been analyzed by the maximum entropy method (MEM) [Collins, 1982]. Figures 2.48(a), (b), (c) & (d) show the 3D view of the unit cell for various Ce concentrations, x=0.02, 0.04, 0.06 & 0.08 respectively. The 3D electronic structure and 2D electron density maps were constructed by considering the similar iso-surface level of 1 $e/Å^3$. Barium, titanium and oxygen atoms and the electron density iso-surfaces around the atoms have been clearly visualized from these figures 2.48(a), (b), (c) & (d). The clear picture of inter-atomic bonding and the charge distribution around them are visualized by constructing 2D electron density contour maps.

The nature of the Ba-O and Ti-O bonds has been analyzed for two most significant crystallographic planes (100) and (200). Figure 2.49 (a) represents the 3D unit cell with the (100) plane shaded. Figures 2.49 (b), (c), (d) & (e) show the 2D electron density contour maps corresponding to the (100) plane with the enlarged view of bonding between the Ba and O atoms for the $BaTi_{1-x}Ce_xO_3$ ceramic solid solutions with x=0.02, 0.04, 0.06 & 0.08 respectively. It can be seen that, there is no accumulation of charge density between the Ba and O atoms which indicates that, the bonding between Ba-O is ionic in nature. For the increased doping levels of Ce, the charge density contours are faded away from the middle portion of the Ba-O bond path. This behavior clearly indicates that the Ba-O bond becomes more ionic with successive Ce^{4+} addition. Figure 2.50(a) shows the 3D unit cell with the (200) plane shaded. Figures 2.50(b), (c), (d) & (e) represent 2D electron density contour maps corresponding to the (200) plane with the enlarged view of bonding between the Ti and O atoms for the $BaTi_{1-x}Ce_xO_3$ ceramic solid solutions with Ce concentrations, x=0.02, 0.04, 0.06 &0.08 respectively. The charge contours between Ti and O are strongly focused along the bond path which results in the overlapping of electron density distribution, and makes the Ti-O bond less ionic. The hybridization between Ti-3d and O-2p states confirms that the bonding in this perovskite ceramic system cannot be purely ionic but also possesses a partial covalent part [Sonali et al., 2000].

To confirm the bonding nature between the atoms quantitatively, the 1D electron density line profiles are drawn and analyzed. Figure 2.51 shows the 1D electron density line profile of the Ba-O bond and figure 2.52 shows the 1D electron density line profile of the Ti-O bond for various Ce concentrations. The bond lengths and electron density values at the bond critical points (BCP) are presented in table 2.18. The bond lengths of the Ba-O and Ti-O bond increase with the increasing doping level of Ce which is in accordant with the XRD results. The electron density at the bond critical point (BCP) is 0.286 e/Å^3 for the Ce concentration x=0.02 which decreases up to 0.257 e/Å^3 with the increasing doping concentrations of Ce. The low electron density value between the Ba and O atoms authenticates the Ba-O bond becoming more ionic in nature with increasing incorporation of Ce. In the case of the Ti-O bond, the electron density value at the bond critical point (BCP) is 0.626 e/Å^3 for the Ce concentration x=0.02 and the electron density value increases up to 0.707 e/Å^3 for the increased dopant levels of Ce. Increasing substitution of Ce (atomic number: 58) at the Ti (atomic number: 22) lattice site increases the density of electrons along the Ti-O bond path thereby reduces the ionic nature along the Ti-O bond.

3.8.5 $Ba_{1-x}Sr_xTi_{0.9}Zr_{0.1}O_3$

The electronic structure and bonding interaction between the constituent atoms in $Ba_{1-x}Sr_xTi_{0.9}Zr_{0.1}O_3$ (Sr&Zr (co-doping)) ceramic systems with four different Sr concentrations x=0.00, 0.05, 0.07, & 0.14 have been analyzed by the maximum entropy method (MEM) [Collins, 1982]. Using the structure factors extracted from the Rietveld refinement [Rietveld, 1969], MEM [Collins, 1982] computations have been carried out.

Figures 2.53(a), (b), (c) & (d) show the 3D unit cell of $Ba_{1-x}Sr_xTi_{0.9}Zr_{0.1}O_3$ corresponding to various Sr doping levels x=0.00, 0.05, 0.07 & 0.14 respectively. The atomic positions of Ba, Ti and O atoms with the electron density around them are clearly visualized from these figures 2.53(a), (b), (c) & (d). The qualitative bonding analysis of the constituent atoms was performed to understand the electronic structure of Sr, Zr doped $BaTiO_3$ by constructing the 2D electron density contour maps. The electronic bonding analysis of Ba-O and Ti-O was carried out by plotting 2D charge density contour maps in the contour range of 0 e/Å^3 to 1 e/Å^3 and with the contour interval of 0.04 e/Å^3 correspond to two different crystallographic planes (100) and (200). Figure 2.54(a) shows the 3D unit cell of BSZT with the (100) plane shaded.

Figures 2.54(b), (c), (d) & (e) demonstrate the 2D electron density contour mapping along with the enlarged Ba-O bonding regions on the (100) plane for different Sr concentration levels of x=0.00, 0.05, 0.07 & 0.14 respectively. There is no charge linkage between the Ba and O atoms, which shows the ionic nature of the bond. Figure 2.55 (a) shows the 3D unit cell of BSZT with the (200) plane shaded. Figures 2.55(b), (c), (d) & (e) show the 2D electron density contour maps with the enlarged bonding regions around the Ti and O atoms on the (200) plane for the Sr doping levels, x=0.00, 0.05, 0.07 & 0.14 respectively. The overlapping of charge density distribution along the Ti-O bond authenticates the covalent nature of the bond.

The quantitative bonding analysis was done to understand the strength and nature of the Ti-O and Ba-O bond. Figure 2.56 shows the 1D electron density line profile of the Ba-O bond and figure 2.57 shows the 1D electron density line profile of the Ti-O bond for various Sr concentrations. The bond lengths and the electron density values at the bond critical points (BCP) of the Ba-O & Ti-O bonds for various Sr concentrations are given in table 2.19. The Ba-O and Ti-O bond lengths are reducing with the increasing incorporation of Sr concentrations which is consistent with the XRD results. For the Ba-O bond, when Sr concentration x=0.00, the electron density value at the bond critical point (BCP) is 0.278 $e/\text{Å}^3$. When the Sr concentration increases, the electron density values increase up to 0.302 $e/\text{Å}^3$. The electron density values evidenced the ionic nature of the Ba-O bond. For the Ti-O bond, the electron density value at the bond critical point (BCP) is 0.761 $e/\text{Å}^3$ with the Sr concentration x=0.00 and when the Sr concentration increases, electron density value decreases to 0.680 $e/\text{Å}^3$. The electron density values between Ti and O ions evidenced the covalent nature of the Ti-O bond.

Table 3.6 gives the comparison of the mid bond electron density values for the Ba-O and Ti-O bonds for the doped systems, $Ba_{1-x}Sr_xTiO_3$, $BaTi_{1-x}Zr_xO_3$, $Ba_{1-x}La_{2x/3}TiO_3$, $BaTi_{1-x}Ce_xO_3$ & $Ba_{1-x}Sr_xTi_{0.9}Zr_{0.1}O_3$. The nature of bonds with the dopant levels (x) of the doped systems with various ions (Sr, Zr, La, Ce and Sr&Zr (co-doping)) are also given in the table 3.6.

The substitution of Sr at the lattice site of Ba increases the ionic nature between Ba & O ions and decreases the ionic nature between Ti & O ions. The doping of Zr at the lattice site of Ti decreases the ionic nature between Ba & O ions and increases the covalent nature between Ti & O ions. The incorporation of La at the lattice site of Ba increases the ionic nature between Ba & O ions and increases the covalent nature between Ti & O ions compared to undoped sample, but these variations are not systematic due to the presence of the A-site cation vacancy. The substitution of Ce at the lattice site of Ti increases the ionic nature between Ba & O ions and increases the covalent nature between Ti & O ions. The substitution of Sr at the lattice site of Ba in Sr & Zr (co-doping) BSZT systems decreases the ionic nature between Ba & O ions and decreases the covalent nature between Ti & O ions.

By comparing the mid bond electron density values of all the doped titanates, in the $BaTi_{1-x}Ce_xO_3$ system, for Ce concentration x=0.06, Ba-O bond has become more ionic with the mid bond density value of 0.257 e/$Å^3$. The Ba-O bond has become less ionic with the mid bond density value of 0.515 e/$Å^3$ for $Ba_{1-x}Sr_xTiO_3$ system with Sr concentration x=0.2. In the $Ba_{1-x}Sr_xTi_{0.9}Zr_{0.1}O_3$ system, for Sr concentration x=0.00, the Ti-O bond has become more covalent with the mid bond density value of 0.761 e/$Å^3$. The Ti-O bond has become less covalent with the mid bond density value of 0.413 e/$Å^3$ for $Ba_{1-x}Sr_xTiO_3$ system with Sr concentration x=0.2.

Table 3.6 *Comparison of mid bond density values and nature of bond for doped titanates.*

Samples	Ba-O		Ti-O	
	Mid bond density (e/$Å^3$)	Nature of bond with x	Mid bond density (e/$Å^3$)	Nature of bond with x
BST	0.453 - 0.515	increase in ionic nature	0.413 - 0.709	decrease in ionic nature
BZT	0.277 - 0.401	decrease in ionic nature	0.657 - 0.738	Increase in covalent nature
BLT	0.278 - 0.380	increase in ionic nature	0.578 - 0.742	increase in covalent nature
BCT	0.257 - 0.286	increase in ionic nature	0.626 - 0.707	Increase in covalent nature
BSZT	0.278 - 0.302	decrease in ionic nature	0.680 - 0.761	decrease in covalent nature

BST - $Ba_{1-x}Sr_xTiO_3$
BZT - $BaTi_{1-x}Zr_xO_3$
BLT - $Ba_{1-x}La_{2x/3}TiO_3$
BCT - $BaTi_{1-x}Ce_xO_3$
BSZT - $Ba_{1-x}Sr_xTi_{0.9}Zr_{0.1}O_3$

References

[1]   Badapanda T, Senthil V, Rout S. K, Cavalcante L. S, Simoes A. Z, Sinha T. P, Panigrahi S, de Jesus M. M, Longo E, Varela J. A, Curr Appl Phys. 11, 1282 (2011). https://doi.org/10.1016/j.cap.2011.03.056

[2]   Balachandran R, Yow H. K, Ong B. H, Anuar K, Teoh W. T, Tan K. B, ICSE
        2008 Proceedings, Johor Bahru, Malaysia (2008).

[3]   Cavalcante L. S, Anicete-Santos M, Sczancoskic J. C, Simoes L. G. P, Santosa M.
        R.M.C, Varelac J.A, Pizanid P.S, Longoc E, J Phys Chem Solids. 69, 1782 (2008).
        https://doi.org/10.1016/j.jpcs.2007.12.022

[4]   Chen, Zhi Y, Zhi J, Ruyan G, Bhalla A. S, Cross L. E, Appl. Phys. Lett. 80, 3424
        (2002) https://doi.org/10.1063/1.1473871

[5]   Chen Y L and Yang S F, Adv appl ceram. 110, 257 (2011).

[6]   Cheng X. B, Liu H. X, Li Z, Yao Z. H, Liu Y, Yu Z.Y, Cao M. H, J Ceram
        Process Res. 9, 576 (2008).

[7]   Collins D. M, Nature. 298, 49 (1982). https://doi.org/10.1038/298049a0

[8]   Cullity B. D, Stock S. R, Elements of X-ray diffraction, Pearson education. 3rd
        edn. Prentice Hall, Upper Saddle River, 558 (2001).

[9]   Dash S K, Kant S, Dalai B, Swain M D and Swain B B. Indian. J. Phys. 88, 129
        (2014). https://doi.org/10.1007/s12648-013-0395-0

[10]   Ganguly M, Rout S. K, Sinha T. P, Sharma S. K, Park H. Y, Ahn C. W, Kim I. W
        J. Alloy. Compd. 579, 473 (2013). https://doi.org/10.1016/j.jallcom.2013.06.104

[11]   Izumi F, Dilanien R. A, Recent Research Developments in Physics Part II, Vol.3,
        Transworld Research Network. Trivandrum, 699–726, (2002).

[12]   Lin M. H, Lu H. Y, Mater.Sci. Eng. A323, 167 (2002).
        https://doi.org/10.1016/S0921-5093(01)01356-9

[13]   Mancic D, Paunovic V, Vijatovic M, Stojanovic B, Zivkovic Lj. Sci Sinter. 40,
        283 (2008). https://doi.org/10.2298/SOS0803283M

[14]   Momma K, Izumi F, VESTA: a three-dimensional visualization system for
        electronic and structural analysis. J. Appl. Crystallogr. 41, 653 (2008).
        https://doi.org/10.1107/S0021889808012016

[15]   Morrison F. D, Coats A. M, Sinclair D. C, West A. R J. Electroceram. 6, 219
        (2001). https://doi.org/10.1023/A:1011400630449

[16]   Petricek V, Dusek M, Palatinus L, Kristallogr Z, Crystallographic Computing
        System JANA 2006: General features, 229, 345 (2014).

[17]   Piskunov S, Heifets E, Eglitis R. I, Borstel G, Comput Mater Sci. 29, 165 (2004).
        https://doi.org/10.1016/j.commatsci.2003.08.036

[18] Ramoska T, Banys J, Sobiestianskas R, Vijatovic Petrovic M, Bobic J, Stojanovic B. Processing and Application of Ceramics. 4, 193 (2010). https://doi.org/10.2298/PAC1003193R

[19] Ridha N. J, Yunus W.M.M, Halim S.A, Talib Z.A, Mohamad Al-Asfoor F.K, Primus W.C, Am. J. Eng. Appl. Sci. 2, 661 (2009). https://doi.org/10.3844/ajeassp.2009.661.664

[20] Rietveld H. M, J. Appl. Crystallogr. 2, 65 (1969). https://doi.org/10.1107/S0021889869006558

[21] Saravanan R, GRAIN software, Personal communication.

[22] Shannon R. D, Acta Crystallogr. 32, 751 (1976). https://doi.org/10.1107/S0567739476001551

[23] Sonali S, Mookerjee T. P, Phys. Rev. B. 62, 8828 (2000). https://doi.org/10.1103/PhysRevB.62.8828

[24] Souza I.A, Gurgel M.F.C, Santos L.P.S, Goes M.S, Cava S, Cilense M, Rosa I.L.V, Paiva-Santos C.O, Longo E. J. Chem. Phy. 322, 343 (2006).

[25] Suasmoro S, Pratapa S, Hartanto D, Setyoko D, Dani U. M, J. Eur. Ceram. Soc. 20, 309 (2000). https://doi.org/10.1016/S0955-2219(99)00143-0

[26] Thongtha A, Angsukased K, Riyamongkol N, Bongkarn T, Ferroelectrics. 403, 68 (2014). https://doi.org/10.1080/00150191003748907

[27] Urek S, Drofenik M, Makovec D, J. Mater. Sci. 35, 895 (2000). https://doi.org/10.1023/A:1004794223988

[28] Vijatovic M. M, Stojanovic B. D, Bobic J. D, Ramoska T, Bowen P Ceram. Int. 36, 1817 (2010). https://doi.org/10.1016/j.ceramint.2010.03.010

[29] Wenhu Y, Yongping P, Xiaolong C, Jinfei W, JPCS. 152, 012040 (2009).

[30] Wodecka-Dus B, Czekaj D, Arch. Metall. Mater. 54, 923 (2009).

[31] Wood D. L, Tauc J, Phys Rev B. 5, 3144 (1972). https://doi.org/10.1103/PhysRevB.5.3144

[32] Wyckoff R.W.G, Crystal structures, Vol.1, Inter-space publishers, London, (1963).

Chapter 4

Conclusion

Abstract

Chapter 4 gives the major conclusion obtained from the structural analysis, average grain size of all doped titanates, optical band gap studies, surface morphological analysis, chemical composition of the prepared ceramic solid solutions, bonding investigation and electron density distribution of various doped barium titanate ceramic materials.

Keywords

Analysis, Average Grain Size, Optical Band Gap, Particle Size, Electron Density, Chemical Composition

Contents

Conclusion

In this research work, five series of doped barium titanate dielectric and ceramic materials,

    (i) $Ba_{1-x}Sr_xTiO_3$

    (ii) $BaTi_{1-x}Zr_xO_3$

    (iii) $Ba_{1-x}La_{2x/3}TiO_3$

    (iv) $BaTi_{1-x}Ce_xO_3$

    (v) $Ba_{1-x}Sr_xTi_{0.9}Zr_{0.1}O_3$

were prepared and investigated using powder X-ray diffraction (PXRD), UV-visible spectrophotometry (UV-vis), scanning electron microscopy (SEM) and energy dispersive X-ray spectroscopy (EDS). The results obtained from various experimental characterization and analytical techniques are analyzed and summarized.

(i)   Structural

Structural analysis of five differently doped titanates was carried out using powder XRD patterns obtained in the 2θ range of $10°$ - $120°$ with the precise step size of $0.02°$. The experimental XRD patterns were subjected to powder profile refinement for structural studies.

The $Ba_{1-x}Sr_xTiO_3$ ceramic solid solutions were prepared by high temperature solid state reaction technique and subjected to XRD analysis. The experimental XRD patterns revealed that, $Ba_{1-x}Sr_xTiO_3$ ceramic solid solutions have been crystallized in single phased cubic perovskite structure with space group $Pm\bar{3}m$. The isovalent substitution of smaller Sr^{2+} at the lattice site of larger Ba^{2+} systematically shifts the XRD peaks towards higher 2θ angles. The decrease in cell parameter and unit cell volume values with the increasing doping levels of Sr, evidenced from the powder profile refinement due to the ionic size differences.

The XRD analysis of $BaTi_{1-x}Zr_xO_3$ ceramic solid solutions prepared by the high temperature solid state reaction technique, confirmed that the synthesized samples have been crystallized in cubic perovskite structure with space group $Pm\bar{3}m$. The isovalent substitution of larger Zr^{4+} at the lattice site of smaller Ti^{4+} shifts the Bragg peaks towards lower 2θ angles. The cell constant and unit cell volume increase with the increasing Zr doping concentrations, evidenced from powder profile refinement which shows a lattice expansion.

The $Ba_{1-x}La_{2x/3}TiO_3$ ceramic samples prepared by the chemical method and subjected to XRD analysis, confirmed that the prepared samples have been crystallized in tetragonal symmetry with space group $P4mm$. The aliovalent rare-earth substitution of smaller La^{3+} at the lattice site of larger Ba^{2+} shifts the Bragg peaks towards higher 2θ angles except for the composition x=0.015, due to slight lattice disorder. The cell constant values and unit cell volume decrease with increasing incorporation of La^{3+} as evidenced by the powder profile refinement.

The XRD analysis of $BaTi_{1-x}Ce_xO_3$ ceramic compositions prepared by the solid state reaction method showed that, the synthesized samples have been crystallized in single phased cubic perovskite structure with space group $Pm\bar{3}m$. The isovalent rare-earth substitution of larger Ce^{4+} at the lattice site of smaller Ti^{4+} systematically shifts the XRD peaks towards lower 2θ angles. An increase in cell constant and unit cell volume is confirmed from the powder profile refinement.

The XRD analysis of $Ba_{1-x}Sr_xTi_{0.9}Zr_{0.1}O_3$ ceramic systems prepared by the solid state reaction method, confirmed that the prepared samples have been crystallized in single phased cubic perovskite structure with space group $Pm\overline{3}m$. The isovalent substitution of smaller Sr^{2+} at the lattice site of larger Ba^{2+} shifts the Bragg peaks towards higher 2θ angles. A decrease in cell constant and unit cell volume values is observed from the powder profile refinement.

Thus, the successful substitution of various dopants in the crystal structure of $BaTiO_3$ perovskite structure has been confirmed qualitatively by XRD peak shifting trend and also quantitatively confirmed by the lattice parameter values obtained from the powder profile refinement.

(ii) Grain size

The average grain size of the coherently diffracting domains for the doped titanates was evaluated using the Scherrer formula through powder XRD data. The average grain size ranges from 22 nm to 36 nm for $Ba_{1-x}Sr_xTiO_3$, 23 nm to 28 nm for $BaTi_{1-x}Zr_xO_3$, 21 nm to 36 nm for $Ba_{1-x}La_{2x/3}TiO_3$, 17 nm to 26 nm for $BaTi_{1-x}Ce_xO_3$ and 21 nm to 32 nm for $Ba_{1-x}Sr_xTi_{0.9}Zr_{0.1}O_3$. After the synthesis, the pellets were forcefully ground for the characterization studies, resulting the grain sizes to be in the nanometer range.

(iii) Optical

The optical band gap of all the doped titanates was analyzed using UV-vis spectrophotometry. Using the Tauc plot methodology, the optical band gap value has been evaluated. The optical band gap decrease with the increasing incorporation of various dopants (Sr, Zr, Ce, Sr & Zr (co-doping)). The reduction in band gap values indicates the slight reduction in the insulating property of the doped titanate systems. However, in $Ba_{1-x}La_{2x/3}TiO_3$ ceramics, no systematic variation in the optical band gap values was observed due to the presence A-site cation vacancy in the La doped solid solutions.

(iv) Morphological

The surface morphology and microstructure of the synthesized ceramic samples were analyzed using scanning electron microscope (SEM) images. When the dopants Sr^{2+} & Zr^{4+} are incorporated in the crystal structure of $BaTiO_3$, particles with irregular sizes and shapes are formed. The La^{3+} substituted ceramic samples, prepared through the chemical method possess dense microstructure which consists of predominantly smaller grains with spherical shape. The fine grained microstructure of fairly narrow size particles with uniform distribution is evident for all the undoped and La doped samples, due to the chemical method of synthesis in which, the growth will be natural and force-free. The microstructure also consists of large voids in between them.

The incorporation of Ce^{4+} in the crystal structure of $BaTiO_3$ leads to the formation of aggregated particles with low porosity. The Sr & Zr (co-doping) ceramic samples possess dense microstructure of $BaTiO_3$. Sr & Zr co-doping promotes good grain growth and also leads to the formation of particles with well developed grain boundaries. Thus, it is evident that, the substitution of various dopants in the $BaTiO_3$ structure uniquely modifies the surface morphology and microstructure of the doped samples.

The particle sizes are also determined from SEM images for all the doped titanates. Smaller sized particles in the range of 0.2 μm to 0.4 μm are formed for La doped samples (BLT) synthesized through the chemical method. The shape of the particles is also uniform in the naturally grown samples (chemical method). The Sr & Zr co-doped $BaTiO_3$ ceramic systems develop the particles with larger size in the range of 5 μm to 9 μm solid solutions prepared through the solid state reaction method. This shows that, the BSZT samples prepared through the high temperature solid state reaction technique promotes good grain growth.

(v)   Elemental

The chemical compositions of the synthesized ceramic samples were thoroughly analyzed by EDS analysis. The presence of each element in each composition of synthesized ceramic solid solutions was confirmed qualitatively by EDS spectra. The stoichiometry of all the doped titanate systems was confirmed by the atomic percentages and mass percentages of each element present in the ceramic systems quantitatively.

(vi)   Charge density

The precise electronic structure, chemical bonding and the electron density distribution between the constituent atoms for all the doped titanates were systematically analyzed by charge density studies.

$BaTiO_3$ perovskite structure has both covalent and ionic character of bonds. The qualitative and quantitative analysis of electron density revealed the predominant ionic nature of the Ba-O bonds and the covalent nature of the Ti-O bond for all the doped titanates.

The systematic variation in the ionic nature along the Ba-O bond and covalent nature along the Ti-O bond has been obtained with the respective dopants in the selected concentration levels, by the analysis of 2D electron density contour maps. The nature of the Ba-O and Ti-O bonds has also been quantitatively confirmed from the 1D electron density line profiles.

Investigation of mid bond electron densities reveals that, in the $BaTi_{1-x}Ce_xO_3$ system, for Ce concentration $x=0.06$, Ba-O bond has become more ionic with the mid bond density value of 0.257 e/Å^3. Ba-O bond has become less ionic with the mid bond density value of 0.515 e/Å^3 for $Ba_{1-x}Sr_xTiO_3$ system with Sr concentration $x=0.2$. In the $Ba_{1-x}Sr_xTi_{0.9}Zr_{0.1}O_3$ system, for Sr concentration $x=0.00$, the Ti-O bond has become more covalent with the mid bond density value of 0.761 e/Å^3. The Ti-O bond has become less covalent with the mid bond density value of 0.413 e/Å^3 for $Ba_{1-x}Sr_xTiO_3$ system with Sr concentration $x=0.2$.

In this research, the charge density distribution and chemical bonding of five different doped barium titanate ceramic systems have been analyzed in detail. These analyses give precise information on the electronic structure of the synthesized samples. The variation in nature of bond and various structural, optical, morphological properties with the influence of dopant elements are compared and analyzed. This detailed analysis of the doped titanate systems facilitates the better understanding of the properties to develop new high performance electronic devices to meet present day technological advancements.

Keyword Index

About the Author

Dr Ramachandran Saravanan, has been associated with the Department of Physics, The Madura College, affiliated with the Madurai Kamaraj University, Madurai, Tamil Nadu, India from the year 2000. He is the head of the Research Centre and PG department of Physics. He worked as a research associate during 1998 at the Institute of Materials Research, Tohoku University, Sendai, Japan and then as a visiting researcher at Centre for Interdisciplinary Research, Tohoku University, Sendai, Japan up to 2000.

Earlier, he was awarded the Senior Research Fellowship by CSIR, New Delhi, India, during Mar. 1991 - Feb.1993; awarded Research Associateship by CSIR, New Delhi, during 1994 – 1997. Then, he was awarded a Research Associateship again by CSIR, New Delhi, during 1997- 1998. Later he was awarded the Matsumae International Foundation Fellowship in1998 (Japan) for doing research at a Japanese Research Institute (not availed by him due to the simultaneous occurrence of other Japanese employment).

He has guided eleven Ph.D. scholars as of 2017, and about five researchers are working under his guidance on various research topics in materials science, crystallography and condensed matter physics. He has published around 140 research articles in reputed Journals, mostly International, apart from around 50 presentations in conferences, seminars and symposia. He has also guided around 60 M.Phil. scholars and an equal number of PG students for their projects. He has attracted government funding in India, in the form of Research Projects. He has completed two CSIR (Council of Scientific and Industrial Research, Govt. of India), one UGC (University Grants Commission, India) and one DRDO (Defense Research and Development Organization, India) research projects successfully and is proposing various projects to Government funding agencies like CSIR, UGC and DST.

He has written 8 books in the form of research monographs including; "Experimental Charge Density - Semiconductors, oxides and fluorides" (ISBN-13: 978-3-8383-8816-8; ISBN-10:3-8383-8816-X), "Experimental Charge Density - Dilute Magnetic Semiconducting (DMS) materials" (ISBN-13: 978-3-8383-9666-8; ISBN-10: 3-8383-9666-9) and "Metal and Alloy Bonding - An Experimental Analysis" (ISBN -13: 978-1-4471-2203-6). He has committed to write several books in the near future.

His expertise includes various experimental activities in crystal growth, materials science, crystallographic, condensed matter physics techniques and tools as in slow evaporation, gel, high temperature melt growth, Bridgman methods, CZ Growth, high vacuum sealing etc. He and his group are familiar with various equipment such as: different types of cameras; Laue, oscillation, powder, precession cameras; Manual 4-

circle X-ray diffractometer, Rigaku 4-circle automatic single crystal diffractometer, AFC-5R and AFC-7R automatic single crystal diffractometers, CAD-4 automatic single crystal diffractometer, crystal pulling instruments, and other crystallographic, material science related instruments. He and his group have sound computational capabilities on different types of computers such as: IBM – PC, Cyber180/830A – Mainframe, SX-4 Supercomputing system – Mainframe. He is familiar with various kind of software related to crystallography and materials science. He has written many computer software programs himself as well. Around twenty of his programs (both DOS and GUI versions) have been included in the SINCRIS software database of the International Union of Crystallography.